The Life of Yeasts

The Life of Yeasts

H. J. PHAFF, M. W. MILLER, AND E. M. MRAK

Second Edition
Revised and Enlarged

HARVARD UNIVERSITY PRESS

CAMBRIDGE, MASSACHUSETTS

AND

LONDON, ENGLAND

1978

Second edition
Printed in the United States of America
Library of Congress Cataloging in Publication Data

Phaff, Herman Jan, 1913–
The life of yeasts.

Bibliography: p.
Includes index.
1. Yeast. I. Miller, Martin Wesley, 1925– joint author. II. Mrak, Emil Marcel, 1901– joint author. III. Title.
QR151.P49 1978 589'.23 77-15108
ISBN 0-674-53325-9

Preface

The preface to the first edition of *The Life of Yeasts* stated that the book was written to serve the nonspecialist who occasionally comes in contact with this group of microorganisms and who wishes to orient himself further without resorting to complex treatises written mainly for specialists. While the second edition has been completely revised and rewritten and a large amount of new information has been added, the purpose expressed in the first edition still holds true.

In the past decade knowledge of yeasts has accumulated far more rapidly than in several previous decades. It became evident, therefore, to the authors and publisher that a revised edition of *The Life of Yeasts* was needed to incorporate this newer knowledge and to make the book more useful and scientifically up-to-date. The developments described in this new edition are not only extensive but very exciting, and they have made the yeasts one of the most interesting groups of eukaryotic microorganisms.

The study of physiological processes has advanced because so many biological chemists use yeasts as a tool. Geneticists also employ yeasts extensively, so there have also been great advances in this area. Other scientists, including nutritionists, microbiologists, agriculturists, food scientists, ecologists—in fact, all those interested in the biological world—have contributed to the study of yeasts. Even taxonomy has progressed during the past ten years, and we now know that the group of organisms we call yeasts includes representatives of not just one class of sexually reproducing fungi, but two.

Most physiologists and biochemists who use yeasts as tools work with baker's yeast or brewer's yeast (*Saccharomyces cerevisiae* or *Saccharomyces carlsbergensis*), perhaps as a result of their availability or because of a lack of knowledge of other yeasts. During the last decade, however, the fascinating diversity revealed in the metabolic, biochemical, nutritional, genetic, and ecological properties of yeasts has encouraged use of a wider spectrum of species for research. In this volume, we have tried to offer a reasonably balanced series of topics and have placed greatest emphasis on the biological aspects.

The first chapter, which deals with the historical aspects of yeast, has not been changed appreciably from the first edition. Chapter 2 discusses the morphology of yeast cells and the manner in which they propagate asexually into certain cell arrangements, an area in which knowledge has expanded enormously in the last few years. This chapter has been enlarged and additional illustrations included.

The third chapter concerns the ultrastructure and function of the various cell components and organelles. The coverage is much more comprehensive, and many illustrations, including electron micrographs, have been added. Here indeed is newer knowledge.

Chapter 4, on the sexual reproduction of yeasts, covers the life cycles not only of the ascomycetous yeasts, but also of the basidiomycetous yeasts, which were not clearly understood when the first edition was published. Additional photomicrographs of sexual spores of both classes are incorporated to give us a more detailed understanding of what yeasts really are.

Although significantly expanded, Chapter 5 remains only an introduction to the highly specialized and complex field of yeast genetics. The enormous literature of the past ten years precludes a more comprehensive coverage.

Chapter 6, concerned with metabolic activities, has been enlarged to cover mainly those aspects unique to yeasts; as in the first edition, the general biochemistry and details of metabolic pathways, which can be found in standard textbooks, receive little emphasis.

In the chapter on nutrition, we discuss normal growth requirements, the unusual nutrients needed by certain species, and special environmental conditions that occasionally may be required for growth. We have included a brief section on the important antibiotics effective against yeasts.

Newer information on specific habitats of yeasts will be found in an updated Chapter 8. The area is a fascinating one, of special interest to biologists concerned with the distribution of yeasts, the conditions under which they can and will grow, and their relation to other organisms. The conditions under which yeasts actually live and propagate in nature, and the type of niche they occupy, are covered in detail. This intriguing area is normally least understood by the nonspecialists.

The spoilage of food by yeasts, a special aspect of ecology, is treated in a separate chapter, which is followed by an account of the uses of yeast in industry. Descriptions of several new substrates (hydrocarbons, methanol, and ethanol) for the production of single-cell protein have been added to this edition, and there is a section on the production of sour-dough French bread.

As is the case with general physiology, taxonomic information cannot be covered adequately in a book this brief. Our treatment of taxonomy, therefore, is limited to elementary principles of yeast classification. We have, however, added some important recent developments in molecular taxonomy, such as DNA base composition and DNA/DNA complementarity.

Finally, the appendixes provide descriptions of the principal genera and an alphabetical listing of the genera in tabular form. The listing will serve as a quick reference for readers who encounter unfamiliar generic names in the various chapters or in the literature. A glossary of mycological terms has been retained for the same reason.

As in the first edition, we have limited the bibliography to books, reviews, and an occasional important reference to an original publication. Although the text might be more useful if we were to document it fully by references to the original literature, the amount of space required to do so would inflate the size of the book undesirably.

As a result of all these changes, the second edition of *The Life of Yeasts* is actually more a new book than a revision of the first edition. The authors hope that the contents will stimulate researchers as well as students to take an interest in the yeasts, a group of microorganisms that has inspired us for many years.

We should like to express our appreciation to Mary Miranda and André Lachance for their help in preparing several of the illustrations. We are indebted to Dr. W. T. Starmer for his critical reading of the entire manuscript and for his constructive advice. Finally, we should like to acknowledge the excellent secretarial help of Sherry Bailey.

Contents

Figures

Tables

The Life of Yeasts

1 / Historical Aspects

The word "yeast" immediately brings to mind the word "fermentation", for the two terms have been closely associated throughout their histories. Although yeast was used before history was recorded, people were in complete ignorance of the nature of these organisms. This is indicated by the use of the term "yeast" in many languages as merely a description of its gross appearance or of what it did for the ancients. The French term for yeast, *levure,* comes from the Latin *levere*, meaning "to raise." This derives from the evolution of carbon dioxide during fermentation, which appears to raise the surface of the liquid as a foam. The German *Hefe* comes from a stem *heben*, also meaning "to lift." The English word "yeast" and the related word "gist" in Dutch are derived from the Greek term *zestos*, which means "boiled"—also a reference to the bubbling foam caused by the evolution of carbon dioxide during fermentation.

The fermentation of fruit juices, and more indirectly the extracts of cereal grains, by mixed populations of yeast probably constituted the first uses of yeast by mankind and resulted in types of alcoholic beverages that may be considered the forerunners of our present-day wine and beer. Even today, certain primitive tribes in Peru are known to pretreat cereal grains by chewing the kernels prior to fermentation to convert starch by salivary amylases into fermentable sugars. It is easy to imagine the early incorporation of fermenting liquid into bread dough and hence the origin of raised, or leavened, bread. Just when these practices of accidental origin were first discovered by humankind is a matter of speculation. That fermentations and bread making were well established 4,000 years ago is graphically shown by models of a brewery and bakery found in an Egyptian tomb at Thebes on the Nile. The exodus of Moses and the Jews from Egypt, as related in the Bible, explains their use of unleavened bread; in their haste to depart from Egypt the leavening was left behind. Apparently all of the ancient civilizations utilized the products of fermentation very much as we do today.

The development of the concept of yeast per se is relatively recent and may be considered to have had its beginning with the recorded observations of van Leeuwenhoek in 1680 of tiny "animalcules." He observed a variety of minute, living things in droplets of various materials with the microscopes he had made as a hobby. The instrument with which he observed these very small bodies actually was not a microscope as we know it, but rather a tiny, carefully hand-ground and polished lens, set between metal plates, which could be focused upon the object by moving the object toward the lens. Van Leeuwenhoek's microscopes were capable of magnifying objects only 250–270 times their natural size. But his extreme skill in grinding and polishing these lenses and his unusual perceptiveness enabled him to discover microorganisms as small as yeasts.

One of the materials he examined was a droplet of fermenting beer. The numerous yeast cells it contained he described as globular bodies, sometimes oval or spherical. These first observations of yeast cells were recorded by van Leeuwenhoek as drawings and as descriptions in letters to members of the Royal Society of London. The significance of his findings, however, was not realized by his contemporaries, nor by those who followed immediately, for it was 150 years later before additional information on yeast was forthcoming. The intervening century and a half was taken up by the experiments of numerous proponents and opponents of the theory which claimed that living things could develop spontaneously.

The theory of spontaneous generation probably existed since people thought about the origin of living things. In the fourth century B.C. Aristotle asserted that animals could arise from different kinds of animals, plants, or even soil. The influence of his concepts was still strongly felt as late as the seventeenth century. Experiments by Francesco Redi and other opponents of this theory helped to quell the idea as far as macroscopic animals were concerned. The discovery of the tiny "animalcules" by van Leeuwenhoek started arguments anew as to their origin. In 1749 Needham claimed they arose from meat, whereas shortly afterward Spallanzani kept microorganisms from appearing in meat by boiling the meat for one hour and then sealing the flask in which it was cooked. Claims that air was vital to the spontaneous generation of microorganisms were subsequently disproved by Schulze and by Schwann in the early nineteenth century and by Schroeder and von Dusch in about 1850. Schulze and Schwann used air that was heated to high temperatures; the latter used cotton plugs to filter the air, which was allowed to enter flasks of heated meat extract. These two treatments prevented the development of microbes in sterilized broth. Such experiments designed to disprove spontaneous generation essentially ended with Pasteur's work in 1864, when he used flasks with long, thin necks curved into a gooseneck shape. In such a flask air could pass into the flask without obstruction, but particles and germs in the air impinged upon the walls of the neck and did not gain access to the broth where they could multiply. Tyndall supplemented these experiments by showing that dust carried germs. If the air was dust free no growth of organisms would occur.

Knowledge gained during the early and middle parts of the nineteenth century served as a basis for several modern fields of biological science. Present-day biochemists, physiologists, and nutritionists all have an interest in the discovery that fermentation is the consequence of yeast's vital activities. The view expressed by Erxleben in 1818 that yeast consisted of living vegetative organisms responsible for fermentation received little attention at that time. New interest in this view was stimulated some years later by Cagniard de la Tour in France (1835) and by Schwann and Kützing (1837) in Germany. These investigators developed the so-called vitalistic theory of fermentation. They proposed that if yeasts are introduced into a sugar-containing solution, they use the sugar as food and excrete the nonutilizable parts as alcohol and carbon dioxide. To the chemists of that period—in particular, von Liebig and Wöhler—this theory was completely unacceptable, and von Liebig countered with his mechanistic theory. His purely chemical fermentation theory left no room for the participation of a living substance. According to von Liebig, yeast is a substance that is continually in the process of chemical transformation. This was supposed to cause the decomposition of dissolved sugar into ethanol and CO_2, because the yeast was thought to impart its "atomic motion" to the sugar molecule. For many years von Liebig's views were generally accepted as the correct ones. It was the result of Pasteur's genius (he, incidentally, was also trained as a chemist) that the vitalistic theory finally triumphed over the mechanistic theory. Pasteur, who had become increasingly involved with biological problems, presented his views on fermentation in integrated form in *Études sur la bière*, published in 1876. Pasteur postulated that the fermentation process for organisms living under anaerobic conditions (including both anaerobic bacteria and yeast) constituted a substitute for the respiratory processes. For example, alcoholic fermentation was considered the vital energy-yielding process of yeast living in the absence of oxygen: "La fermentation est la vie sans air."

Pasteur's demonstration that yeast possesses a respiratory besides a fermentative ability, and his experiments, which tended to show that aeration repressed fermentation, rounded out his studies on the physiological activities of yeast. This fundamental aspect of yeast metabolism, later studied in a more quantitative manner by Meyerhof, is now usually referred to as the "reaction of Pasteur–Meyerhof" (See Chapter 6).

A partial return to the mechanistic theory occurred near the turn of the century as a result of the accidental discovery of the "zymase" of yeast in 1897 by Eduard and Hans Buchner. Many, including the great French investigator of yeasts A. Guilliermond, considered this finding as the downfall of the fundamental part of Pasteur's fermentation theory. The Buchner brothers attempted to prepare an extract from brewer's yeast for medicinal purposes. They ground yeast with diatomaceous earth and squeezed out the juice in a mechanical press. They added sugar to the cell-free juice as a preservative and were startled to find that the juice began to bubble and froth. The process that took place was a cell-free alcoholic fermentation, and the agent responsible for the fermentation was termed zymase, now known to consist of a complex mixture of enzymes. In fact, the word "enzyme" is derived from the Greek word meaning "in yeast."

This discovery reduced the role of the yeast plant to the simple secretion of zymase. Consequently, many chemists of that period thought Buchner's discovery was the final victory of von Liebig's mechanistic theory. They considered the "dead" zymase as a chemical or catalytic agent responsible for splitting sugar into alcohol and CO_2.

Subsequent investigations, however, have shown that the cell-free fermentation is quite unbalanced as compared with the process that takes place inside the living cell. The significance of fermentation to the living cell is that fermentation (dissimilation or catabolism) of sugar yields energy that enables the cell to utilize parts of the sugar molecule for growth (assimilation or anabolism). These reactions, in the living cytoplasm of the cell, have reestablished the vitalistic theory of fermentation, which, with minor modifications, is the generally accepted one today.

Following the discovery of zymase many biochemists of the twentieth century—including Neuberg, Meyerhof, Warburg, Wieland, Embden, Parnass, Harden, and others—contributed to the complete elucidation of the intermediary pathway and energetics of the fermentation process. The significance of their discoveries was greatly magnified when these same reactions (often referred to as glycolysis) were shown to take place in many different types of living cells and to represent a basic process by which an organism obtains (from sugar) energy for life.

After van Leeuwenhoek's description of yeast in 1680, very little additional information on its morphology was gained until the early nineteenth century.

Starting about 1825, Cagniard de la Tour, Kützing, and Schwann, in their studies on beer and wine yeasts, showed that these organisms are cells that reproduce by budding. A fermenting liquid was diluted until tiny droplets contained very few cells: the microscopic reproduction of the cells could be seen under a microscope. To prevent drying out during the period of observation, the droplets were placed on a coverslip of thin glass, which in turn was placed upside down on a glass slide with a hollowed-out area in the surface (a moist chamber), and the edges of the coverslip were sealed with wax or paraffin. Yeast cells in the droplets could then be observed during growth.

In 1839 Schwann observed "endospores," bodies formed within yeast cells. He also noted the freeing of these internal bodies by rupture of the cell wall, but the significance of these spores was not understood.

De Bary, in 1866, compared the spore-containing bodies of yeast with the spore sac of ascospore-forming fungi termed *Ascomycetes.* Reess, in a series of experiments between 1868 and 1870, observed endospores in many species of yeasts and gave accurate descriptions of their shapes and their manner of germination by budding. He termed these endospores "ascospores" and the spore sac an "ascus," since he too noted the similarity in spore development with certain of the lower *Ascomycetes*. In 1870 he suggested the name *Saccharomyces* (previously used by Meyen in 1837, who first used the term for budding yeasts) for the spore-forming yeasts and included them in the *Ascomycetes*.

Later, in 1884, de Bary pointed out that yeast spores as well as ascospores formed by the more complex *Ascomycetes* are produced by "free cell formation"—that is, the spore bodies are delimited within the protoplasm of the ascus, or spore sac, free from other spores and also free from attachment to the ascus wall. This is in contrast to the method of spore formation in other classes of fungi—notably the *Phycomycetes (Zygomycotina)* and the *Basidiomycetes (Basidiomycotina)*.

Among his many contributions to science, Pasteur, introduced methods for obtaining pure cultures and thereby made possible more reliable morphological studies of yeasts. The great Danish student of yeast Emil Christian Hansen perfected these techniques, and for thirty years he carefully studied morphological and certain physiological characteristics of yeasts. He was able to differentiate and characterize a large number of species, many of which are still recognized today. He attempted the first comprehensive system of yeast taxonomy in 1896. His career of meticulous investigations earned him recognition as the true founder of morphological studies of yeasts.

Hansen's recommendations of systematic relationships and life cycles of yeasts were expanded by Guilliermond in Paris, who contributed much additional information on the physiology, sexuality, and phylogenetic relations among the yeasts in his monographs of 1920 and 1928. Guilliermond also devised various dichotomous keys for use in identifying yeast species.

After Guilliermond five monographs on the taxonomy of the major groups of yeasts emerged between 1931 and 1970 from the Technical University of Delft. These works, although they did not carry his name, were inspired directly or indirectly by the great Kluyver of the "Delft School." The first complete and usable scheme of classification for the sporulating yeasts was prepared by Stelling-Dekker in 1931. Subsequently, Lodder in 1934 and Diddens and Lodder in 1941 published two volumes on the non-spore-forming yeasts. The publication of these three studies brought some unity and order out of chaos in the field of yeast taxonomy and greatly facilitated the classification of yeast cultures by the average microbiologist. A comprehensive classification of both the sporogenous and asporogenous yeasts was published in 1952 by Lodder and Kreger-van Rij. This volume represented a careful examination and reevaluation of 1317 strains of yeast maintained in the Yeast Division of the Centraalbureau voor Schimmelcultures, housed in the Laboratory for Microbiology of the Technical University at Delft, Holland. These strains were classified into 165 species with 17 varieties. At about the same time (1951) Wickerham in the United States introduced a number of novel techniques and principles in the classification of yeasts. For example, he introduced several synthetic media to study morphology, assimilation of a much more representative array of carbon compounds, and ability to grow with or without vitamins in such media (see Chapter 7). He also placed greater emphasis on the occurence of yeasts in nature in the haploid or in the diploid form and on the existence of heterothallic mating types (Chapters 4 and 5 give further details.)

A monograph comprising a rather different system of yeast taxonomy was published by the Russian taxonomist Kudriavzev in 1954. His scheme is interesting because of some of his views on phylogeny of the ascospore-forming yeasts.

Through the cooperative efforts of fourteen taxonomists in various countries of the world, the classification of the yeasts was further revised and updated in 1970 under the editorship of J. Lodder. Since the monograph by Lodder and Kreger-van Rij in 1952, the number of species described in the new monograph increased from 165 to 341. Three years later Barnett and Pankhurst completed a novel identification key, based primarily on biochemical properties. They included not only the 341 species described in the 1970 monograph, but 93 additional species described between 1970 and 1973—a total of 434. Discoveries by present-day investigators continually add new species, so that the total today exceeds 500, classified in some 50 genera (see also chapter 11).

The historical developments sketched here have so far offered little information about just what a yeast is. Early investigators in the field, working with typical beer and wine yeasts, found these to be generally unicellular organisms that reproduced by budding and formed ascospores under suitable conditions within a mother cell or ascus. The origin of the word "yeast" related primarily to its ability to ferment. Definitions applied to yeast according to their industrial behavior, such as "cultured, true, wild, top or bottom yeasts," have little meaning from a botanical point of view. They are confusing even from the industrial point of view, for a cultivated yeast considered desirable in a brewery may be considered a wild yeast in a bakery. Top-fermenting brewery yeasts generally rise to the surface of the liquid as a foam and should have the ability to split bond (a) of the trisaccharide raffinose and to ferment only the fructose portion, since they are unable to hydrolyze the residual melibiose.

$$\underbrace{\text{Galactose}\overset{\text{(b)}}{-\alpha-(1\rightarrow 6)-}\text{Glucose}}_{\text{Melibiose}}\overset{\text{Sucrose}}{\overbrace{-\alpha-(1\rightarrow 2)-\beta-\text{Fructose}}}\ \text{(a)}$$

Bottom-brewing yeasts, on the other hand, should split both bonds (a) and (b) and ferment the entire raffinose molecule. In addition, the cells of a bottom yeast tend to settle to the bottom of the liquid soon after fermentation subsides. Though this differentiation is seemingly clear-cut, such is not the case; there is no provision for forms that are intermediate in their sedimentation behavior, and these are known to exist.

From a botanical point of view, yeasts are indeed difficult to define as a single homogeneous group. A discussion of a "mushroom," or perhaps a "fern," would permit a listing of rather specific characteristics that all mushrooms or all ferns would have in common.

Such specific characteristics encompassing all yeasts are difficult to designate. Nineteenth-century botanists generally accepted the idea that yeasts belong to the plant kingdom. Today many taxonomists agree with the arrangement proposed by Ainsworth in which the fungi are treated as a separate kingdom. In this scheme the yeasts are included in the division *Eumycota*.

Yeasts lack chlorophyll and are unable to manufacture their organic needs by photosynthesis from inorganic components as do higher plants, algae, and even some bacteria. Therefore, they must live a saprophytic or parasitic life. Yeasts possess thick, rather rigid cell walls, have a well-organized nucleus with a nuclear membrane (eukaryotic), but lack any means of locomotion. These properties fit the characteristics of the division *Eumycota*, while the additional abilities of forming sexual spores within an ascus or producing them externally on a basidium place some yeasts in the subdivision *Ascomycotina* and some in the *Basidiomycotina*, respectively. Those yeast species in which a perfect (sexual) stage is not known are included in the subdivision *Deuteromycotina*.

Yeasts are usually defined as fungi that do not produce asexual spores (conidia) born on distinct aerial structures (conidiophores) and that spend at least part of the vegetative cycle as single cells. Budding, while a very common means of asexual reproduction, is not the only manner of vegetative reproduction by yeasts. Certain species reproduce vegetatively by cross-wall formation (sometimes followed by fission), and still others by a process intermediate between budding and cross-wall formation (bud fission). Another group has the ability to form asexual spores externally on the tip of pointed, specialized structures termed "sterigmata." These spores are forcefully discharged (by a mechanism still not fully understood). For this reason they are termed "ballistospores" (See Chapter 4). This last group represents yeasts that appear related to the *Basidiomycotina*.

As with most biological materials, there are transitional forms between yeasts and the more typical higher fungi, in which case the organisms in question are usually designated as yeast-like fungi. To complicate the definition of yeast even more, some unicellular algae produce natural mutants devoid of chlorophyll. These colorless algae, common in nature, are probably derived from *Chlorella* and are normally placed in the genus *Prototheca*. Unlike yeasts, their vegetative reproduction involves neither budding nor fission, but a kind of internal cell partitioning into an indefinite number of daughter cells (aplanospores or spherules). Nevertheless, they can be and have been mistaken for yeasts, since their macroscopic appearance typically resembles yeasts.

Some fungi that belong to the subdivision *Zygomycotina*—in particular, some species of *Mucor*—grow aerobically as mycelial molds on solid media. However, under strictly anaerobic conditions of growth, particularly with carbon dioxide in the atmosphere, these fungi produce budding, yeast-like cells. When transferred back to aerobic conditions, however, these cells promptly revert to the normal mycelial form.

This brief account of the historical development of knowledge about yeasts indicates that these organisms have long been the object of scientific attention. Yet it is still difficult to define accurately what yeasts are. As a group they form a heterogeneous association of organisms with scientifically fascinating properties.

2 / Yeast Morphology and Vegetative Reproduction

As first observed by van Leeuwenhoek and later studied by Cagniard de la Tour, Kützing, and Schwann, yeast cells found in fermenting beverages appeared to be round or oval and to reproduce by forming buds that develop into daughter cells. The new cells separated from the mother cell and produced buds of their own. This description also fits the appearance of many other yeasts. Upon closer scrutiny, however, the shapes and methods of reproduction among the "yeasts" are quite varied.

Vegetative Reproduction

Besides the development of sexual reproductive structures (see Chapter 4), morphological properties are, in general, based on the characteristics of vegetative re-

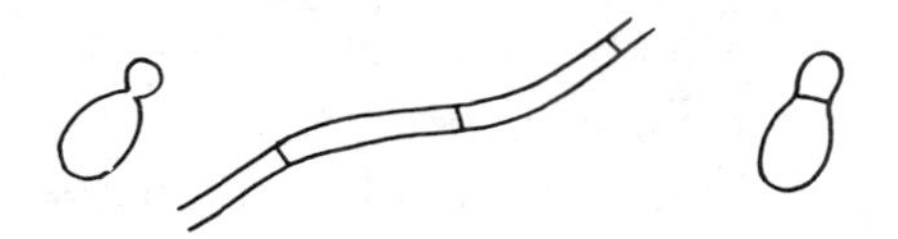

FIG. 1. Cell division by budding, cross-wall formation (septation), and bud fission.

production of a particular yeast (Fig. 1). Budding, as mentioned earlier, represents the most common mode of reproduction. With the exception of a few genera discussed later, buds usually arise on the shoulders and at the ends of the long axes of ovoidal or elongate vegetative cells—thus in areas of greatest cell curvature. This type of budding, which is never repeated at the same site, is referred to as "multilateral" and is characteristic of *Saccharomyces* and most other ascomycetous genera. Where the axes are of similar lengths, such as in spherical cells (often found in species of *Debaryomyces*), buds apparently arise any place on the surface of the cells. Yeasts related to *Basidiomycotina* (for example, *Rhodotorula*) also reproduce by budding. In these yeasts, however, a number of buds may arise at the same site, although some investigators have observed more than a single budding site on the cell surface. In contrast to the multilaterally budding yeasts, there are several genera in the *Ascomycotina* (for example, *Hanseniaspora* and *Saccharomycodes*) where budding is restricted to the poles or tips of the cells (bipolar budding). Such organisms are known as "apiculate" or "lemon-shaped" yeasts; the apiculate shape results from repeated bipolar budding (see Chapter 3). Cells of *Pityrosporum* reproduce by repeated unipolar budding on a broad base and in the genus *Trigonopsis* budding is restricted to the three apices of their triangularly shaped cells.

Species of two genera, *Endomyces* and *Schizosaccharomyces* (Fig. 2a, 2h), reproduce in a vegetative manner exclusively by fission. In this case, reproduction is carried out by the formation of a septum or cross wall, without any constriction of the original cell wall. When the process is complete, the cross wall divides into two individual walls, and the newly formed cells can then separate.

Occasionally a type of vegetative reproduction occurs that is intermediate between budding and cross-wall formation. This so-called bud fission results from a type of budding in which we find a very broad neck at the base of the bud somewhat resembling a bowling pin. Subsequent formation of a septum across the isthmus separates the bud from the mother cell. Examples of this type of reproduction may be found in *Saccharomycodes* (Fig. 2b), *Nadsonia*, and in *Pityrosporum* (the latter with unipolar budding). This type of reproduction is not fundamentally different from that occurring in a typical budding yeast. It is largely a difference in size of the cross wall formed, which in budding yeasts [such as *Saccharomyces* (Fig. 2c)] is so small that it gives the impression under the light microscope that the bud is "pinched off" (see also Chapter 3).

THE THALLUS

The thallus, or soma, is the vegetative body or structure of a yeast plant. The manner in which the component cells of such a structure are arranged may be very characteristic for certain species or even genera.

In its simplest form the thallus may be a single cell or perhaps one with its daughter cell still attached (such as in *Saccharomyces cerevisiae*). Here the first bud usually detaches as soon as the mother cell produces a second bud. In other yeasts, however, buds may remain attached to the mother cell, and the first daughter cell—as well as the mother cell—can produce additional buds, and so on. This may result in small to large clusters or chains of cells (Fig. 2f).

Chain formation can lead to a structure called a "pseudomycelium" (false mycelium). A pseudomycelium (Fig. 2e, 2g) arises when the bud stays attached to the mother cell, elongates, and continues to bud in turn instead of breaking away upon maturity. Often the area widens where two cells join, giving the impression of a true cross wall. In this manner, strands are formed that resemble true mycelium in appearance (cells separated by a cross wall), differing only in the way in which the new cells arise (budding).

Development of pseudomycelia ranges from very rudimentary in which the number of cells is limited and where there is little or no differentiation among the cells, to those in which there is marked differentiation between the cells comprising the main stem (usually composed of rather elongated cells) and the buds arising in clusters on the shoulders of these elon-

a

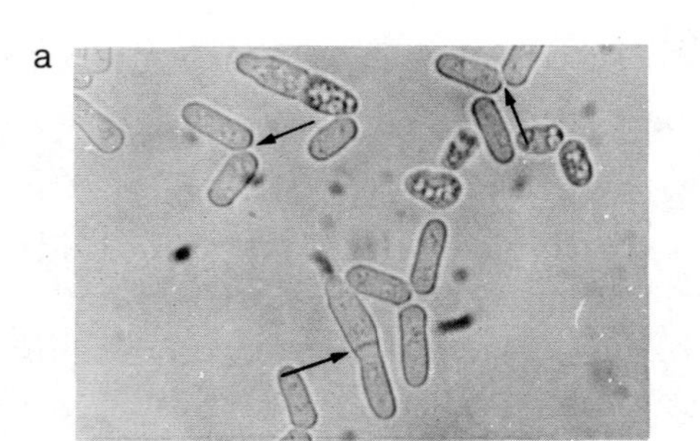

b

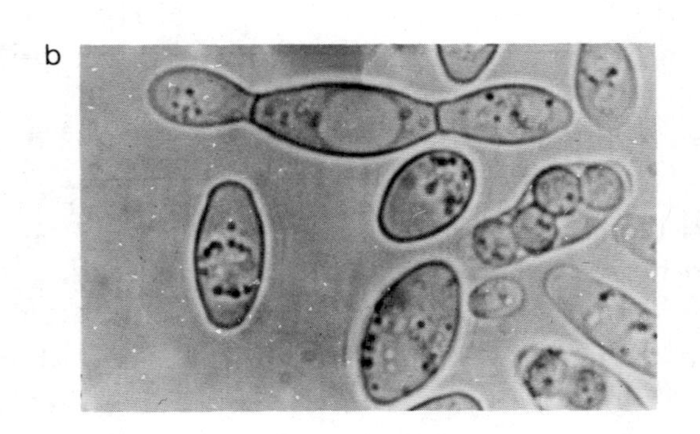

c

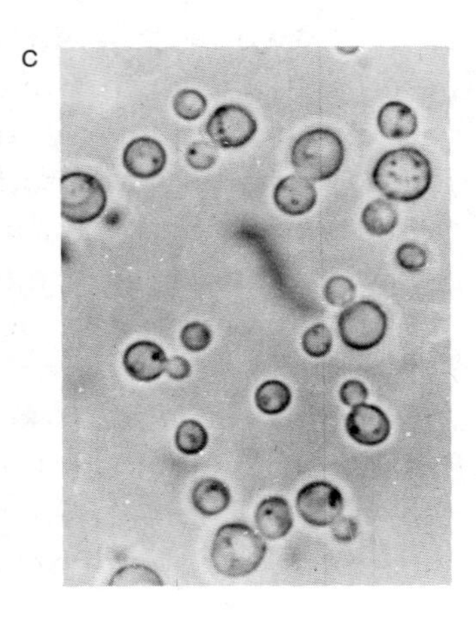

d

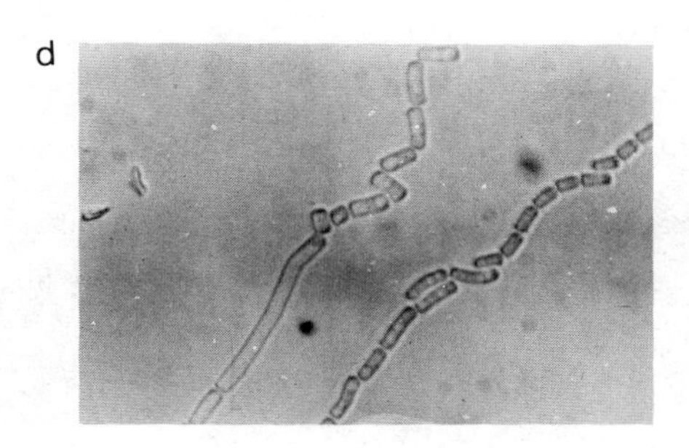

e

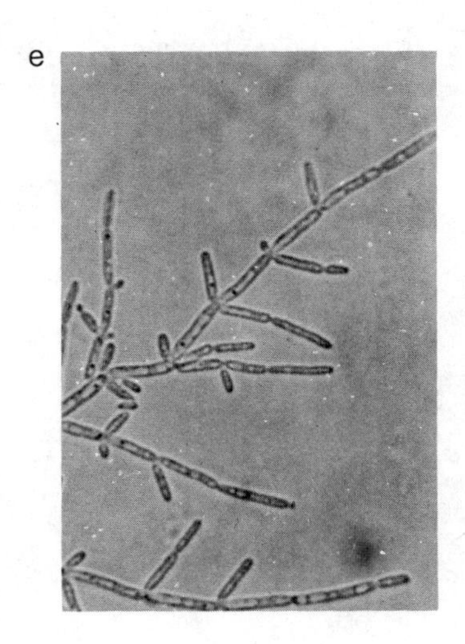

FIG. 2. Vegetative reproduction of yeasts: (*a*) cells reproducing by cross-wall formation (*Schizosaccharomyces*); (*b*) bipolar budding and separation of buds by cross-wall formation; "bud fission" (*Saccharomycodes*); (*c*) budding cells (*Saccharomyces*); (*d*) mycelial hyphae breaking up into arthrospores (*Trichosporon*); (*e*) pseudomycelium formation, simple type nearly devoid of blastospores (*Candida mesenterica*).

f

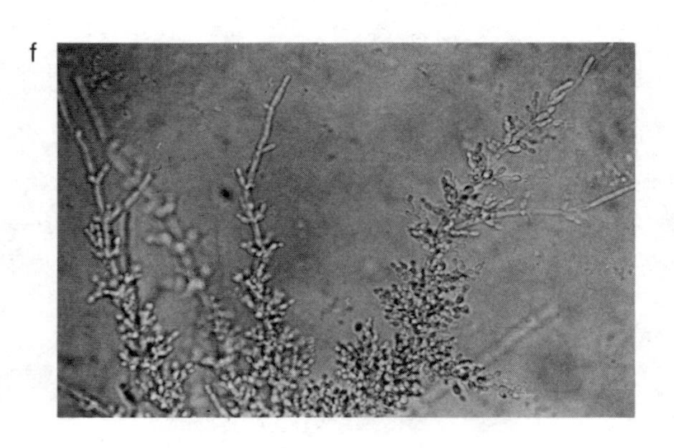

g

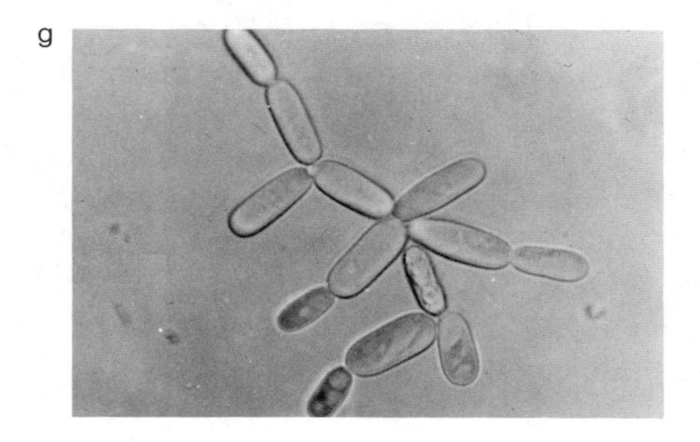

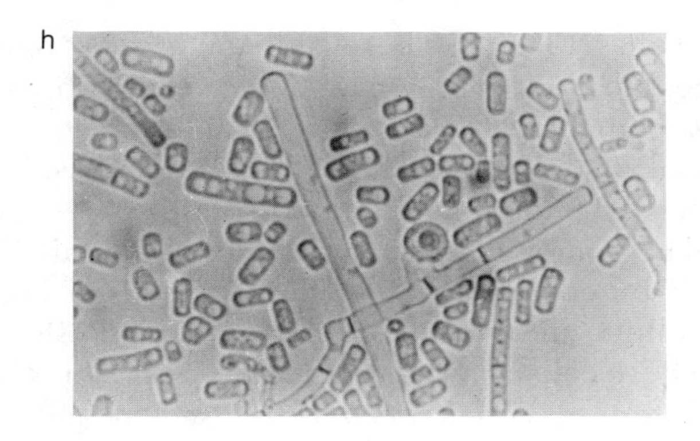

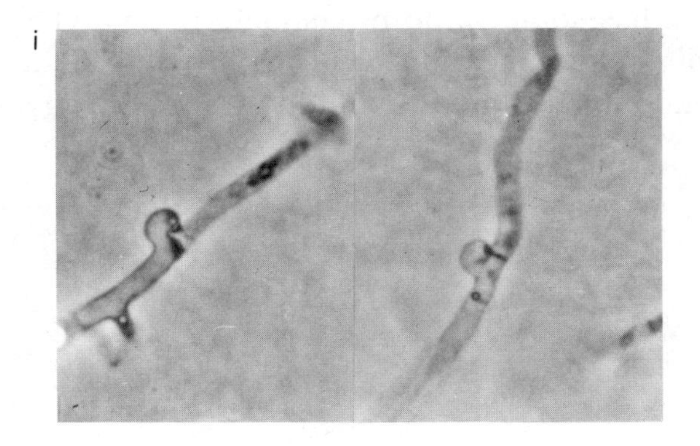

FIG. 2 (*continued*). (*f*) Cluster of budding cells (*Cyniclomyces guttulatus*); (*g*) Pseudomycelium with abundant development of blastospores (*Candida tropicalis*); (*h*) true mycelium and arthrospores (*Endomyces reessii*), note typical ascus formation by gametangial copulation and single ascospore; (*i*) cell division with clamp connections in basidiomycetous yeasts (*Sporidiobolus*).

gated cells (Fig. 3). These side buds may remain spherical to oval or they may elongate and stimulate further branching. As a means of differentiating the spherical or ovoid buds from the pseudomycelial or stem cells, the former are termed "blastospores" (Fig. 2g). Blastospores that are not formed terminally, but are distributed laterally along the stem cells of a pseudomycelium, are sometimes referred to as "blastoconidia." These cells may be attached by short stems. The French term *appareil sporifère* is used by some authors for a well-differentiated pseudomycelium. Even though a given species may produce a characteristic form of pseudomycelium, very often several types of pseudomycelia may be found within a single species. Thus although the ability or inability to form pseudomycelia has some value in characterizing a yeast, the type of pseudomycelium is of lesser value. Examples of pseudomycelia are illustrated in Figs. 2e, 2g, and 3.

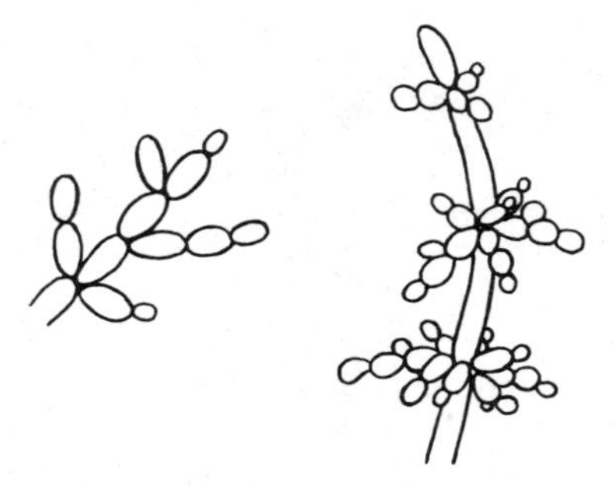

FIG. 3. *Left*, a rudimentary pseudomycelium; *right*, a characteristic branch of a well-developed pseudomycelium.

Yeasts that utilize both budding and fission as means for vegetative reproduction also exist. Members of the genus *Trichosporon* (Fig. 2d) usually grow in the form of hyphal strands, containing cross walls at regular intervals. These strands can undergo disarticulation, which is when they break up along the cross walls into individual cells termed "arthrospores"(or "oidia"). On a solid medium, arthrospores are often seen in zigzag formation that somewhat resembles derailed railroad cars (Fig. 2d). In addition, various budding cells arise on the mycelial strands. In *Saccharomycopsis* (= *Endomycopsis*) and certain species of *Candida* hyphae with cross walls and bud formation are found. But in these two genera the hyphae generally do not break up into arthrospores. In *Saccharomycopsis* the cross walls may have pores that form a connection between adjacent cells (see Chapter 3). Besides budding cells and hyphae with cross walls, some of the basidiomycetous yeasts (for example, *Sporidiobolus* and *Rhodosporidium*) form another type of mycelium having clamp connections between adjoining cells (Fig. 2i). This mycelial type is formed during the dikaryotic phase of their life cycle (Chapter 4).

CELL MORPHOLOGY

Besides spherical, globose, ovoid, elongated, and cylindrical vegetative cells, which may arise as a result of the modes of vegetative reproduction discussed earlier, there are certain yeasts with highly characteristic cell shapes (Fig. 4). The lemon-shaped or apiculate

FIG. 4. Various cell shapes and modes of reproduction. *Left to right:* spheroidal, ovoidal, elongate, ogival, triangular, flask shaped, apiculate, curved.

vegegative cell is characteristic of yeasts commonly found in the early stages of natural fermentation or spoilage of fruits and berries (*Hanseniaspora* and its imperfect form *Kloeckera*). An ogival cell shape, where an elongated cell is rounded at one extremity and somewhat pointed at the other, is characteristic of yeasts named *Dekkera* (imperfect-*Brettanomyces*), which have been used in the past for the production of ale in Ireland, Great Britain, and Belgium. Species of these genera are also found as spoilage yeasts of bottled wines and soft drinks. The "bottle bacillus," associated with dandruff, is a yeast with a flask-like cell shape. This yeast *(Pityrosporum)* reproduces vegetatively by repeated bud fission at one of the cell's poles, and thus gives rise to a bottle or flask shape. The triangular cells of *Trigonopsis* are unique. The single species of this genus has been isolated from beer in a Munich brewery and from grapes. Many of the cells of *Cryptococcus cereanus*, a species occurring in the fermenting juice of rotting cacti, are highly curved.

To say that a particular shape is characteristic of a given species or genus does not mean that every cell in a population will be of that shape. However, it will appear in that form at some period in the ontogenetic development of each cell. "Ontogeny" is the term referring to the history of development of an individual organism. For example, the apiculate (lemon-shaped) yeasts generally start as spherical or ovoid buds that separate from the mother cell and in turn develop buds themselves (Fig. 5). Since budding is bipolar, a young ovoid cell developing buds at its extremities then becomes lemon-shaped in appearance. Because of repeated bipolar budding, older cells may assume a variety of shapes, although in old age an irregular elongated shape is common (See Chapter 3).

Some yeast cells may be 2–3 micrometers (μm) in length, whereas in some species cells may attain lengths of 20–50μm. The width is less variable and usually ranges between 1–10μm. The actual size of the vegetative cells in a young culture may be quite uniform in some species or extremely heterogeneous in size and shape in others. This disparity may be used to differentiate between species or, in some instances, between varieties of the same species. Observations

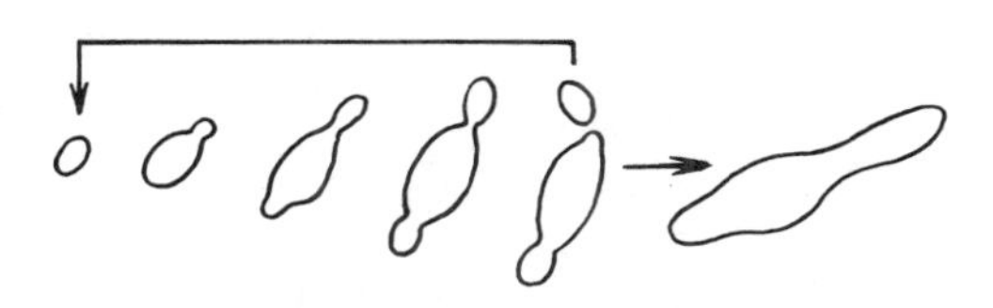

FIG. 5. Development of characteristic cell shapes in an apiculate yeast (*Hanseniaspora*). This is an example of ontogeny—the development of an individual.

on the size and shape of vegetative cells as well as the manner in which they reproduce vegetatively are made both in liquid media and on solid media. Very frequently malt extract is used, although synthetic media have been recommended for this purpose. The latter are more reproducible in their composition. Pseudo- and true mycelium development is usually followed on corn meal agar or potato-glucose agar, since these media stimulate their development. While a certain amount of variation is normal and to be expected within a yeast culture, age, environmental, and cultural conditions can exert a profound influence on the culture's morphological properties. Consequently, descriptions of the various yeast species are based on results obtained under fairly standardized conditions.

3 / Ultrastructure and Function

In discussing the ultrastructure and cellular organization of yeasts, one must guard against generalizations because various species may show significant differences. Much of our earlier information on cytology is based on work with *Saccharomyces cerevisiae*, or baker's yeast. More recently, however, investigators have become interested in species belonging to different genera to bring out particular facets that are very striking in some yeasts and lacking in others. Investigators have also come to recognize that significant cytological changes may be brought about by the cultural conditions under which a yeast is grown prior to study. For example, the thickness of the cell wall, shape of the cell, presence of lipid globules, vacuoles, and inclusion bodies, mitochondrial development, and the extent of capsular polysaccharide formation can be strongly modified by the growing conditions (Fig. 6).

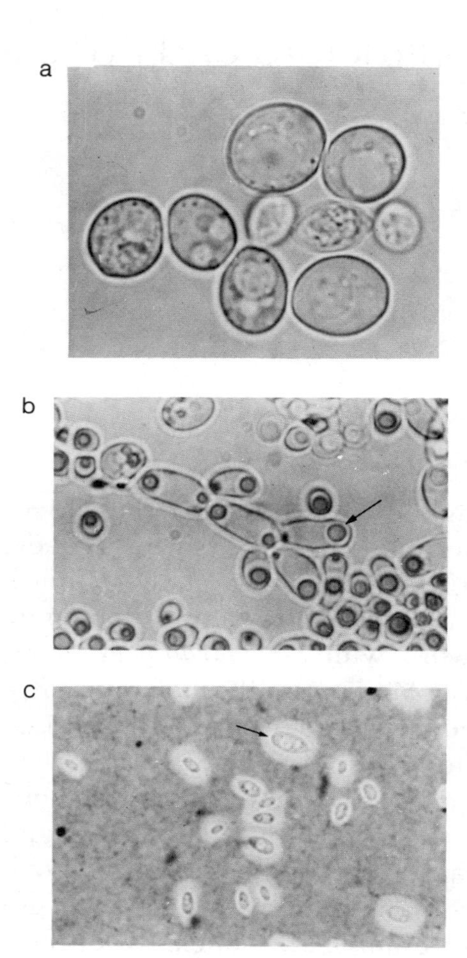

FIG. 6. Structures of yeast cells: (*a*) vacuoles (*Saccharomyces*); (*b*) lipid (oil) globules (*Hansenula*); (*c*) capsules photographed against a background of India ink. The India ink particles do not penetrate the capsular material, resulting in light halos (*Cryptococcus*).

Information on the cytology of yeast has been gathered by:

1. Direct observation with the light microscope.
2. Staining the cell with dyes or brighteners to determine the location of specific components or surface areas.
3. Transmission electron microscopy (TEM) of ultrathin sections of yeast cells or of platinum-carbon replicas of freeze-fractured cells. The latter technique involves extremely rapid freezing of a very concentrated yeast suspension, followed by fracturing the frozen droplet. This fine splintering exposes the surfaces of a number of cell organelles and membranes. The frozen preparation is then etched (that is, the cut surface is freeze-dried to a depth of about 20 nm) and coated with a platinum-carbon replica in high vacuum. After appropriate cleaning, the replica is observed in the transmission electron microscope. The technique reveals surface structures of internal cell organelles with great fidelity.
4. Scanning electron microscopy (SEM) of vegetative structures and ascospores. Because of its great depth of observation, this technique has furnished important information on the topography of vegetative cells and spores as well as details of asexual and sexual reproduction.

The microstructures of a yeast cell are the cell wall (including a mucous capsule in some species), cytoplasmic membrane or plasmalemma, nucleus, one or more vacuoles, mitochondria, microbodies, ribosomes, endoplasmic reticulum, dictyosomes, lipid globules, volutin or polyphosphate bodies, and the cytoplasmic matrix.

THE CELL WALL

The wall can be seen with the oridinary light microscope as a distinct outline of the cell. Although it is somewhat elastic, its rigidity is responsible for the particular shape a yeast cell assumes. Under the light microscope—even with the highest magnification possible—the cell wall does not reveal distinct features and appears as a fairly smooth outline of the cell occasionally dotted with slight irregularities (Figs. 2 and 6).

Much more information on the physical appearance of yeast cell walls has been obtained by the use of electron microscopy and the possibility of preparing isolated walls. When whole cells of a young yeast culture that reproduces by multilateral budding (for example, *Saccharomyces*) are observed in the scanning electron microscope, the most conspicuous features of the seemingly smooth walls are (a) a single birth scar and (b) a variable number of bud scars. A birth scar is a structure on a daughter cell resulting from its separation from the mother cell by the budding process. A birth scar normally occurs at one end of the long axis of an ovoidal cell and does not have a highly characteristic appearance (Fig. 7a). Bud scars are formed when the new cell, in turn, produces daughter cells by budding. Bud scars in ascomycetous, multilaterally budding yeasts show a circular brim, slightly raised above

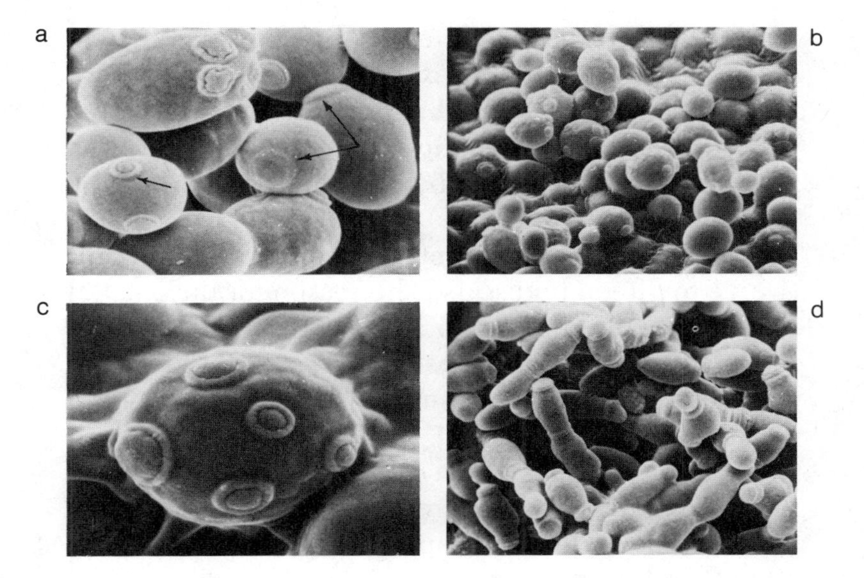

FIG. 7. Scanning electron micrographs of yeasts during vegetative reproduction: (*a*) bud scars (single arrow) and birth scars (double arrows) in a multilaterially budding yeast (*Saccharomyces*); (*b*) random distribution of bud scars (*Saccharomyces cerevisiae*); (*c*) a senescent cell with many bud scars (*S. cerevisiae*); (*d*) annular bud scars characteristic of yeasts that reproduce bipolarly (*Saccharomycodes*).

the cell surface, surrounding a central area of about 3 μm^2. The number of bud scars depends on how many daughter cells a mother cell has produced, because in such yeasts successive buds never form at the same site (Fig. 7b).

The number of buds a yeast can form is limited. Investigators have determined that strains of *Saccharomyces cerevisiae* can produce as many as nine to forty-three buds per cell, with a median of twenty-four. In these experiments successive buds were removed with the aid of a micromanipulator and cell division was not limited by exhaustion of nutrients. One should realize, however, that the great majority of the cells in a normal population which has reached the stationary phase of growth have either no bud scars (unbudded daughter cells) or very few (one to six). Only a minute fraction of the cells carry perhaps twelve to fifteen bud scars. The reason for the smaller than maximal number of bud scars on the oldest fraction of the population is cessation of growth due to exhaustion of a particular nutrient and/or crowding of cells.

In diploid species of *Saccharomyces* the pattern of scar distribution is more or less random (Fig. 7b). In haploid *Saccharomyces* species the bud scars are often found in rows, rings, or spirals. After the production of three or four daughter cells the mother cell gradually assumes a different appearance, and surface irregularities can sometimes be seen, even with the light microscope. Under ideal conditions a yeast cell may duplicate in a time period between one and a half and two hours, but after many buds have been produced the generation time lengthens and ultimately may become as long as six hours. Finally, the cell stops budding and presumably dies. The cell wall of old cells may assume a distinctly wrinkled appearance (Fig. 7c), and cell death could be the result of an impairment in the uptake of nutrients.

In ascomycetous yeasts with apiculate cells, where budding occurs exclusively at the two poles, the bud scars are superimposed on each other, giving a peculiar type of scar tissue characterized by a series of ringlike ridges on the polar extensions of the cell (Fig. 7d). The more buds a cell has produced, the longer the polar extensions of an apiculate cell become.

In yeasts related to the basidiomycetous fungi (for example, species in the genus *Rhodotorula*), repeated bud formation at the same site often occurs. The walls of successive buds arise each time underneath the original cell wall, giving rise to concentric collars. These can be best demonstrated in a longitudinal ultrathin section through the bud scar and viewed by TEM (Fig. 8). Thus, very few bud scars are seen on the cell surface. In addition, the scars are less striking and the circular ridge is less well developed than in ascomycetous yeasts when whole cells are viewed by SEM.

In the fission yeasts of the genus *Schizosaccharomyces*, another type of scar is formed. Longitudinal thin sections through cells of *Schizosaccharomyces* viewed by TEM show that cross-wall formation occurs by an annular centripetal growth of an inner lateral

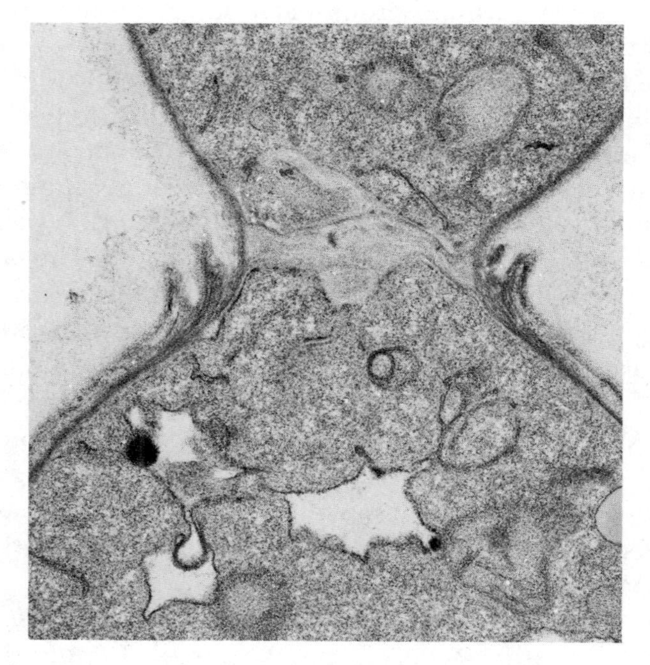

FIG. 8. Transmission electron micrograph of an ultrathin section through a budding basidiomycetous yeast (*Phaffia rhodozyma*); note collar-like remnants of cell wall formed by repeated budding at the same site.

portion of the cell wall. After completion of the cross wall it splits transversely, resulting in two cells. During this cleavage, division scars are formed (as shown schematically in Fig. 9a). During subsequent growth the scar plugs grow out, retaining a circular scar between the original cell and the new growth (Fig. 9b). Upon subsequent fission in the elongated mother cell, in the new outgrowth, or in both, the process repeats itself and additional scar rings (also referred to as plug-wall bands) are formed on the original cell.

Electron microscopy of filamentous, ascomycetous yeasts has shown that the septa separating the cells are not continuous in all species, but many contain pores. The septum normally contains a central pore, surrounded by the swollen edge of the cross wall, and the pore is closed on both ends by a plug. These pore structures have been referred to as "dolipores," and have been found in *Ambrosiozyma* (synonym *Endomycopsis*) *platypodis*, among others. In other species the cross walls are traversed by a number of thin strands of protoplasm, the plasmodesmata. These structures have been found in *Dipodascus aggregatus* and *Saccharomycopsis* (Synonym *Endomycopsis*) *fibuligera*. Finally a single, small central connection has been observed in the septa of other species—for example, *Saccharomycopsis* (*Endomycopsis*) *lipolytica*. These three types of pores are schematically illustrated in Fig. 10.

Our present understanding of the chemical composition of the cell wall of yeasts is based largely on the combined evidence of chemical analysis and enzymatic degradation of the walls. Data have become much more meaningful since clean, purified cell walls, free of cytoplasmic inclusions, became available. Again, most of the work has been done with species of *Saccharomyces*, although more recently other yeasts have been investigated as well.

The *Saccharomyces* cell wall is thought to consist of three layers which probably are merged together in

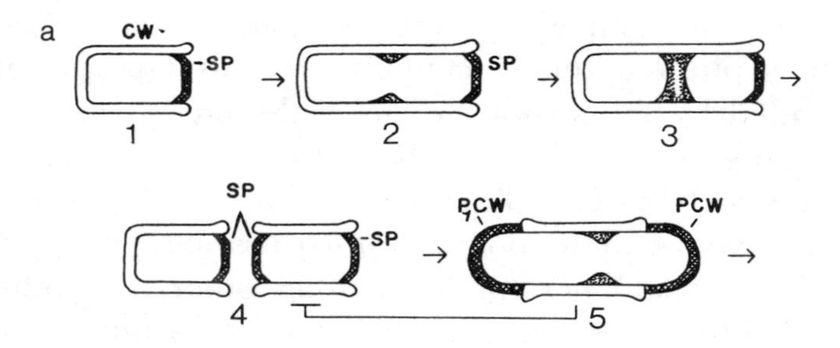

FIG. 9. Cell division in *Schizosaccharomyces*: (*a*) schematic representation of the fission process: (*1*) cell with a single birth scar or scar plug (SP); (*2*) cell elongation and beginning of septum formation; (*3*) septum formation is complete and it begins to divide along its length (light area); (*4*) cell separation is complete, resulting in one cell with a single and one with two scar plugs; (*5*) outgrowth of the scar plugs resulting in two new polar cell walls (*PCW*). Note that this mechanism of reproduction results in ridges on the outer cell wall (*CW*). The fission process then repeats itself (redrawn after E. Streiblova, J. Bacteriol. *91*, 434, 1966). (*b*) Scanning electron micrograph of *Schizosaccharomyces octosporus* showing the scar plugs and cell wall ridges resulting from cell reproduction.

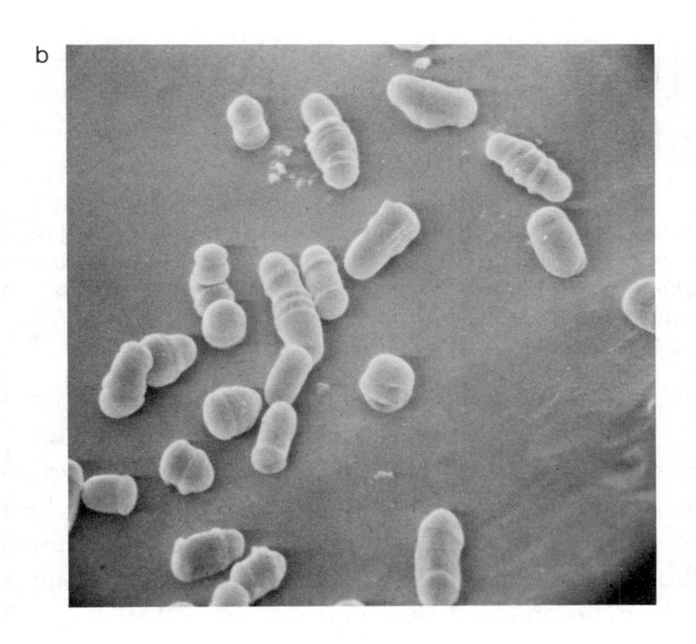
b

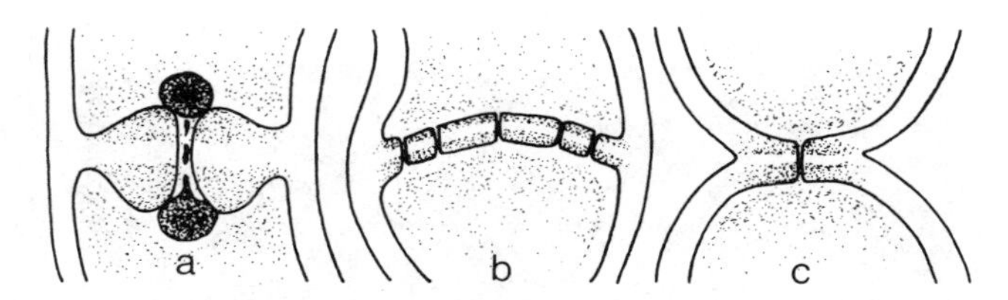

FIG. 10. Schematic representation of various types of septal pores in filamentous ascomycetous yeasts: (*a*) dolipore (*Ambrosiozyma platypodis*); (*b*) plasmodesmata (*Saccharomycopis fibuligera*); (*c*) simple central pore (*Saccharomycopsis lipolytica*).

part: (a) an inner layer of alkali-insoluble β–glucan, (b) a middle layer of alkali-soluble β-glucan, and (c) an outer layer of glycoprotein in which the carbohydrate consists of phosphorylated mannan. The precise location of lipoidal material in the cell wall is not known, although some of it may be accounted for by remnants of plasmalemma not removed during cell wall purification.

When cell walls are extracted with hot alkali, the insoluble residue (about 30–35%) constitutes the glucan component responsible for the shape and rigidity of the yeast cell. Only in recent years have investigators realized that this component represents a mixture of two polysaccharides, a major component (85%) of a predominantly β–(1→3)-linked glucan [with 3% of β–(1→6)-linkages as branch points], and a minor component (15%) of a highly branched β–(1→6)-glucan with β–(1→3)-interchain or interresidue linkages. The minor component can be removed from the major component by numerous extractions with 0.5 M acetic acid at 90°C. The two components do not appear to be covalently linked, and the reason for their tight association is not well understood. There is electron microscopic evidence that the alkali-insoluble glucans occur in the wall as a tightly structured microfibrillar network. The reason for the insolubility of the major component and its ability to form microfibrils with a considerable degree of rigidity is thought to be due to the linking together of β–(1→3)-linked chains of glucose residues by hydrogen bonding and/or by other forces of attraction in triple helices.

The alkali extract of the cell walls consists of a mixture of alkali-soluble glucan and the mannan–protein complex. The latter can be precipated by adding Fehling solution. When the remaining solution is neutralized with acetic acid, a gel-like precipitate is formed of the alkali-soluble glucan. It can be washed with distilled water and represents approximately 20–22% of the cell wall. The main difference in chemical structure between it and the alkali-insoluble glucan is that a number of β–(1→6)-linked glucose residues are present in the long sequences of β–(1→3)-linked glucose chains. Investigators think that these anomalous linkages interfere with hydrogen bonding of neighboring molecules and with the formation of microfibrils. As a result this amorphous glucan has become soluble in alkali. Since this glucan has been found to contain a small proportion of mannose, it is possible that the amorphous glucan in the cell wall is linked to the outer mannan–protein layer and that alkali ruptures some part of the connection between the two layers.

As mentioned earlier the mannan–protein complex (about 30% of the wall) can be precipitated from the alkaline cell wall extract with Fehling solution. After washing with distilled water the copper complex can be dissolved in acid and the mannan precipitated with alcohol. Investigators have found that hot alkali depolymerizes the mannan–protein complex and that a more native mannan is obtained by autoclaving cell walls or whole yeast with a dilute sodium citrate buffer at pH 7.0. Baker's yeast mannan is composed of D–mannose residues that are arranged in the form of an α–(1→6)-linked backbone about 50 units in length, to which are attached short side chains of α–(1→2)- and α–(1→3)-linked mannose residues. The lengths of the side chains vary in different yeast species. There may also be various side groups attached to the side chains—for example, phosphate (in some strains of *Saccharomyces*), mannosyl–phosphate (in *Kloeckera brevis*) and *N*–acetylglucosamine (in some species of *Kluyveromyces*). The side chains in their various configurations (especially the terminal and side groups) elicit specific antibodies in animals after injection with a particular strain of yeast. These side chains with their side groups are, therefore, the antigenic determinants of the yeast cell. That portion of the mannan molecule discussed so far is termed the "outer chain." It is linked to an inner core of about twelve to seventeen mannose residues in a branched configuration, which, in turn, is attached through a β–(1→4) bond to *N*–acetylchitobiose. The latter is linked to asparagine in the protein portion of the molecule. The protein or peptide may carry a number of mannan molecules, each one attached to an asparagine residue. In addition, small mannose oligosaccharides are glycosidically linked to hydroxyl groups of serine and threonine of the peptide chain. The significance of these oligosaccharides is not clear at present.

The protein content of the manno-protein complex which coats the cell wall ranges from about 5–10%. However, mannan–protein enzymes (for example, invertase and acid phosphatase) that contain much more protein (30–50%) are found in the periplasmic space and perhaps intermixed with the other cell wall components. The size of the structural cell wall mannan–protein complex has been determined for the yeast *Kloeckera brevis*. Mannan extracted by hot citrate buffer was resolved into three molecular sizes of 25,000, 100,000 and 500,000 daltons. The smaller fragments may be the result of partial breakdown of native mannan by the alkaline Fehling solution used for precipitating the mannan. When the two larger mannan fragments were treated with the proteolytic enzyme Pronase, their molecular weights were reduced to about 25,000, whereas the smaller component was not affected. These experiments led to the idea of a mannan subunit—in this case a molecule with about 150 mannose residues and 15 amino acids. However, there are important variations in mannan structure between different yeast species and even different strains of the same species (for example, *Saccharomyces cerevisiae*).

Few details are as yet known about the biosynthetic steps involved in the synthesis of this complicated mannan macromolecule. We know, however, that guanosine diphosphate mannose (GDPM) is the mannosyl donor for the carbohydrate portion of the molecule. The mannosyl residues of GDPM can be transferred directly by several highly specific mannosyl transferases to build the main chain and side chains of the mannan molecule. Depending on the enzyme, the mannose units are attached by α–(1→2), α–(1→3), or α–(1→6) bonds. Genes have been identified by preparing recessive mutants lacking one or more of the specific transferases. However, the attachment of oligosaccharides to serine and threonine of the glycoprotein peptide chain requires a lipophilic intermediate, dolichol monophosphate, a molecule containing 14 to 18 isoprenol units. There is some evidence that the newly synthesized mannan molecule is transported across the plasmalemma in small vesicles. Presumably the protein molecule is synthesized first, since in the presence of cycloheximide (an antibiotic that blocks cytoplasmic protein synthesis), mannan synthesis is also inhibited.

Saccharomyces cell walls contain approximately 1% chitin, a linear polymer of *N*–acetylglucosamine in which the constituent monomeric units are linked by β–(1→4) bonds. In early literature the chitin content of yeast cell walls was derived from determinations of their *N*–acetylglucosamine content. Since we did not know then that this amino sugar also occurs in the mannan fraction, early reports on chitin content have little value. We now know that chitin of baker's yeast walls occurs only in the bud scars. When a bud has fully developed, it is first separated from the mother cell by a primary septum consisting of chitin. The primary septum is then covered with glucan and mannan. An unbudded daughter cell, of course, has either no chitin or only traces, whereas a cell with many bud scars may contain a substantial amount. The donor of *N*–acetylglucosamine in chitin biosynthesis is uridine diphosphate *N*–acetylglucosamine. Cabib and coworkers have found that the enzyme chitin synthetase is present in yeast in an inactive form (zymogen), which is activated by one of the yeast proteases at the time the primary septum is deposited during bud formation. In nonbudding yeast the proteolytic activating factor is maintained in an inactive form by a small protein molecule of about 8500 daltons. The exact mechanism of triggering the activation of chitin synthetase at a specific site and time has not been elucidated as yet.

In filamentous yeasts (for example, *Endomyces*) and in species of *Nadsonia, Rhodotorula, Cryptococcus,* and *Sporobolomyces* the chitin content is much higher. There it is probably an actual cell wall component. In contrast, chitin appears to be lacking entirely in species of the fission yeast *Schizosaccharomyces*.

Another glucan wall component has been identified in recent years—pseudonigeran, or α–(1→3)–glucan. This linear polysaccharide, which is soluble in alkali but precipitates upon neutralization, is present as a major wall component in only one ascomycetous genus (*Schizosaccharomyces*), but is much more common in species of some genera related to the basidiomycetes, such as *Cryptococcus*. This polysaccharide is also common in many genera of the higher fungi.

Wall material of baker's yeast has been reported by some to contain from 8.5–13.5% of lipid material, although lower figures have been reported by others. A factor that is likely to influence the lipid content of yeast cell walls is the method used to disrupt whole yeast cells and the resulting retention or removal of the adhering plasmalemma.

CAPSULAR MATERIALS

In discussing the cell walls of yeast, we must also mention the presence of capsular materials (Fig. 6c) and other extracellular substances formed by some yeasts. The principal categories in this connection are phosphomannans, β-linked mannans, heteropolysaccharides (containing more than one type of sugar unit), and hydrophobic substances belonging to the sphingolipid type of compounds.

Certain species of *Hansenula* (such as *H. capsulata* and *H. holstii*, which in their natural habitat are associated with bark beetles that attack coniferous trees), and those of the closely related genera *Pichia* and *Pachysolen* produce extracellular phosphomannans. These slimy, viscous polymers are water soluble and form a sticky layer on the surface of the cells. They contain only the sugar D–mannose and phosphate, the latter linked as a diester. Oligosaccharides with three to five mannose residues are built into a polymer by way of phosphodiester bridges between carbon–1 of one oligosaccharide and carbon–6 at the nonreducing end in the next oligosaccharide in the polymer. The molar ratio of mannose to phosphate is characteristic of the species producing the phosphomannan, although this ratio is also affected by the cultural conditions under which the yeast is grown. In media with suboptimal phosphate levels the mannose:phosphate ratio increases appreciably.

$$[M^1 \longrightarrow {}^2M^1 \longrightarrow {}^3M^1 \longrightarrow {}^3M^1 \longrightarrow {}^2M^1\!-\!O\!-\!\underset{\underset{O^-}{|}}{\overset{\overset{O}{\|}}{P}}\!-\!O\!-\!{}^6M^1 \longrightarrow {}^3M^1 \longrightarrow {}^3M^1 \longrightarrow {}^3M^1 \longrightarrow {}^2M^1\!-\!O\!-\!\underset{\underset{O^-}{|}}{\overset{\overset{O}{\|}}{P}}\!-\!O]_n$$

Proposed structure of the capsular phosphomannan from *Hansenula holstii.* M = α–D–mannopyranosyl residue.

$$[-O\!-\!\underset{\underset{O^-}{|}}{\overset{\overset{O}{\uparrow}}{P}}\!-\!O \longrightarrow {}^6M^1 \xrightarrow{\beta} {}^2M^1\!-\!]_4 \overset{\alpha}{-} O\!-\!\underset{\underset{O^-}{|}}{\overset{\overset{O}{\uparrow}}{P}}\!-\!O\!-\!{}^6\underset{\underset{M^1}{\overset{\uparrow 2}{|\,\alpha}}}{M}{}^1 \xrightarrow{\beta} {}^2M^1 \overset{\alpha}{-} O -$$

Proposed structure of the capsular phosphomannan from *Hansenula capsulata.* M = mannosyl residue of the indicated configuration.

In several slimy, capsulated species of the red yeasts belonging to the genus *Rhodotorula*, the capsular material consists of a linear or slightly branched mannan with alternating β–(1→3) and β–(1→4) linkages. A similar polysaccharide was identified in the nonpigmented yeast *Torulopsis ingeniosa*.

$$—M\xrightarrow{1\ \beta\ 4}M\xrightarrow{1\ \beta\ 3}M\xrightarrow{1\ \beta\ 4}M\xrightarrow{1\ \beta\ 3}M—$$

Proposed structure of the capsular mannan from species of *Rhodotorula*. M = β–D–mannosyl residue.

Among the heteropolysaccharides, the capsular material of *Cryptococcus* species has received the greatest attention. This interest results partially from the antigenic properties of the capsular material of the pathogenic yeast *Cryptococcus neoformans*. Three serotypes have been identified in strains of this species. The nucleus of the capsular polysaccharide consists of an α–(1→2)-linked backbone of mannose residues, to which are attached single D–xylose units and D–mannose residues, each by α–(1→4) bonds. The latter, in turn, carry an α–(1→4)-linked D–glucuronic acid residue, which accounts for the term "acidic heteropolysaccharide." Among minor components, variable proportions of D–galactose and acetyl groups have been reported. There is evidence that the ratios of these various components may vary significantly among different species and strains of *Cryptococcus*.

Numerous other heteropolysaccharides have been identified from other yeasts, such as a polysaccharide from *Lipomyces lipofer* with a D–mannose to D–glucuronic acid ratio of 2:1. Some species of *Lipomyces*, such as *L. kanonenkoae*, also contain D–galactose. The heteropolysaccharide of *Candida bogoriensis* contains glucuronic acid, fucose, rhamnose, mannose, and galactose; that of *Trichosporon cutaneum* is a pentosyl mannan containing both D–xylose and L–arabinose.

Although much of the capsular polysaccharides or slimes adheres to the cell, a large proportion may be released into the medium of growth, particularly in agitated cultures.

Several of the so-called advanced, film-forming species of *Hansenula* (in particular, *H. ciferrii*) form on their surface tetraacetyl phytosphingosine and triacetyl dihydrosphingosine. These are complex hydrophobic compounds that appear to be responsible, at least in part, for the tendency of these yeasts to form pellicles in liquid media and for the matte appearance of their colonies. (Chapter 6 describes the structure of these compounds.)

FIMBRIAE

Recently, hair-like structures termed "fungal fimbriae" were observed on the cell surface of several basidiomycetous and ascomycetous yeasts. In the former these fibrils may be as long as 10 μm, while in the latter group they are only about 0.1 μm in length. These fimbriae, which appear to be proteinaceous, are 5–7 nm in diameter. In *Saccharomyces* these short fibrils appear to be somehow involved in yeast flocculation.

THE CYTOPLASMIC MEMBRANE OR PLASMALEMMA

This structure is located directly beneath the cell wall and plays an important role in the selective entry of nutrients from the growth medium. It also protects the cell from losing low molecular weight compounds by leakage from the cytoplasm and it represents the matrix upon which the cell wall components are deposited during growth of the cell.

The plasmalemma in ultrathin sections is seen in the transmission electron microscope as a unit membrane—that is, a structure about 8 nm thick in which two electron dense (dark) layers are separated by an electron transparent (light) layer. The dark layers are thought to represent proteins that are separated by a layer of lipids and phospholipids. The center of the middle layer is thought to consist of the nonpolar groups and its borders of polar groups of phospholipids. In addition the central layer is thought to contain proteins involved in the entry and exit of solutes and in enzymatic action. Some carbohydrate (mainly mannan) has also been identified. The most prominent enzyme in the membrane is a Mg^{2+}-dependent ATPase, which is active at neutral pH and insensitive to the antibiotic oligomycin. This enzyme is thought to play a role in the active (energy-dependent) transport of certain solutes into the cell.

In thin sections of cells the plasmalemma shows rather deep and numerous invaginations (up to 50 nm deep). In freeze-etched preparations (after removal of the cell wall by lytic enzymes) the plasmalemma is seen sculptured by invaginations that are about 30 nm wide and of variable lengths. In addition, by this technique the outer surface shows a particulate structure. The particles, which were found to be composed of mannan and protein, have a diameter of about 15 nm and are usually found in hexagonal arrays with a lattice period of 18 nm. The particles are thought to move through the membrane and may be involved in formation of the mannan–protein complex of the wall or in the formation of fimbriae.

We can demonstrate the plasmalemma in whole cells suspended in a medium with an osmotic pressure equivalent to that of the cytoplasm by treating them with a lytic enzyme preparation (for example, the digestive fluid of the garden snail or a bacterial preparation). The cell wall dissolves, leaving the cell surrounded with a very thin membrane (the plasmalemma). The cell (or protoplast) then assumes a spherical shape. Upon dilution of the osmotic stabilizer, the cell takes up water, expands, and finally bursts. Membrane fragments can then be collected by a special process of centrifugation. Other investigators prefer to rupture whole cells by a gentle procedure followed by enzymic digestion of the cell walls adhering to the plasmalemma. Chemical analysis of the plasmalemma has revealed a very complex mixture of neutral lipids (mono- , di-, and triglycerides), free and esterified sterols (mainly ergosterol), a complex sphingolipid, glycerophosphatides (primarily phosphatidyl choline, phosphatidyl ethanolamine, and phosphatidyl inositol), and neutral as well as acidic glycolipid. The latter contains a galactose sulphatide that is responsible for its acidic properties. The fatty acids in the various components are long chained (C_{16}–C_{26}), with oleic acid ($C_{18:1}$) being the most abundant fatty acid.

THE NUCLEUS

The yeast nucleus is not easily seen in the light microscope when the cells are mounted in an aqueous suspension. However phase contrast microscopy of yeast growing aerobically on a nutrient medium containing 18–21% gelatin reveals the nucleus. It is composed of an optically dense crescent-shaped portion (the nucleolus) and a more translucent portion that contains the chromatin material. Besides DNA, the nucleus contains various species of RNA and a type of polyphosphate with a chain length of twenty to forty orthophosphate residues. This species of polyphosphate can be extracted from cells with saturated sodium perchlorate in the cold. It appears to accumulate during RNA synthesis, but a definite function for it in the nucleus has not been established.

During budding of ascomycetous yeasts, mitosis is accompanied by elongation and constriction of the nucleus, which normally takes place in the neck of the bud. Both the nucleolus and the chromatin-containing portion divide as observed under phase contrast microscopy and are incorporated in the two daughter nuclei.

Similar observations have been made in stained preparations, except that nuclear constriction is less evident. Specific dyes are available that selectively stain the nucleolus (acid fuchsin and iron haematoxylin) and the chromatinic portion of the nucleus (aceto-orcein and Giemsa solution). Staining with dyes that have an affinity to the chromatin of the nucleus has also revealed a small, heavily stained "lateral granule" or spindle plaque in resting nuclei that expands into an intranuclear fiber in dividing nuclei.

Ultrathin sections of nonproliferating cells viewed by TEM (Fig. 11) show the nucleus as a more or less spheroidal organelle surrounded by a pair of unit membranes (as defined in the section on the plasmalemma). The nuclear envelope is studded with circular pores approximately 85 nm in diameter, and the pores appear to be filled with small granules that may represent subunits of the ribosomes. There is evidence that the nucleolus is the site of synthesis of RNA components of ribosomes. A surface view of the nuclear pores and their distribution can be obtained in freeze-etched preparations of fractured whole cells.

In ascomycetous yeasts the nuclear envelope remains intact during cell division. Thus both mitosis and the division of the nucleolus are intranuclear in such species. Conversely, in basidiomycetous yeasts mitosis is not intranuclear. Basically, the chromatin portion of the nucleus moves into the bud before division, and the nucleolus in the greatly elongated nucleus remains in the mother cell. In the bud a partial breakdown of the nuclear envelope takes place, and the chromatin divides within remnants of the nuclear membrane. The nucleolus appears to dis-

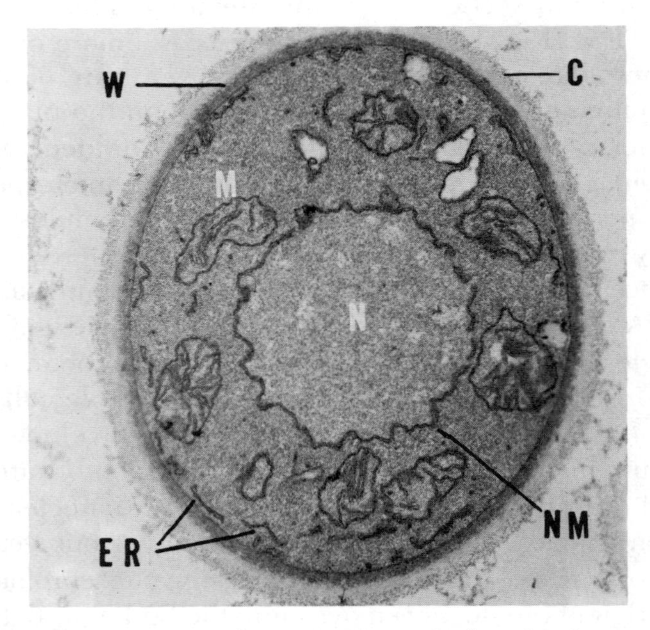

FIG. 11. Transmission electron micrograph of an ultrathin section of a nonproliferating cell (*Phaffia rhodozyma*): (*C* =) capsule; (*ER* =) endoplasmic reticulum; (*M* =) mitochondrion showing cristae; (*N* =) nucleus; (*NM* =) nuclear envelope showing nuclear pores.

integrate as a distinct structure. After chromatin divison, the nuclear envelope reforms around the daughter nuclei, each containing a newly formed nucleolus, and one daughter nucleus moves back into the mother cell. This type of nuclear behavior has been demonstrated in species of *Leucosporidium, Rhodosporidium, Rhodotorula,* and *Sporobolomyces.* The occurrence of mitosis in the bud is considered a useful taxonomic indicator of basidiomycetous affinity in asporogenous yeasts.

In resting (nonbudding) cells of *Saccharomyces cerevisiae*, the G1 interval of the vegetative cell cycle, the nuclear envelope carries a single disk-like structure, the centriolar plaque, from which short microtubules arise. At the end of the G1 interval, three events begin: spindle plaque duplication, the initiation of DNA replication necessary for mitosis, and the emergence of a bud. The interval following G1 has been termed the "S period" or "DNA synthetic interval." It constitutes about 25% of the cell cycle. At the beginning of the S period, the duplicated spindle plaques move away from each other along the nuclear envelope. The two plaques are the site of origin of hollow microtubules 15–18 nm in diameter. After the plaques have separated 180°, a bundle of approximately fifteen microtubules forms a straight continuous fiber connecting the two spindle plaques inside the nucleus. Outside the nucleus, the microtubules extend into the cytoplasm. Since the double plaque (directly after plaque division) is usually located in close proximity to the bud initial and the extranuclear microtubules extend into the new bud, some investigators have suggested that there is a causative relationship between the double plaque and bud initiation. Following the S period is the G2 interval, during which the bud grows in size and the nucleus migrates to the neck of the mother cell where it undergoes the first stage of nuclear division—that is, a medial nuclear constriction, coupled with elongation of the microtubules. In the following M interval the microtubules continue to elongate. The M interval is succeeded by the second stage of late nuclear division, in which separation of the daughter nuclei is completed. This is followed by cytokinesis (separation of the plasmalemma of mother cell and bud) and cell wall separation, completing the cycle with the production of two unbudded cells. Under optimal conditions of growth, a complete cell cycle takes place in about one and one-half hours. (Fig. 12 depicts the various events schematically.) Currently, many researchers are using a genetic approach in attempts to elucidate the factors in the cell that initiate and control the various events in the cycle. Although the microtubules probably participate in the separation of chromosomes, the exact mechanism is not known. In some stained preparations chromosome-like structures have been reported. But under most conditions of staining and fixation both for light and electron microscopy, the chromatin appears diffuse throughout the nucleus. As a result there is no general agreement that yeast nuclei

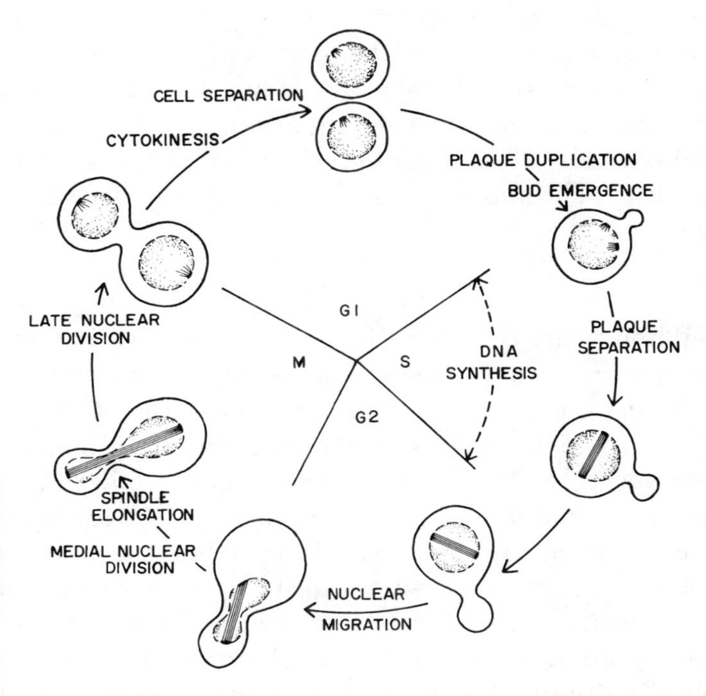

FIG. 12. Diagrammatic representation of bud formation and nuclear behavior during the cell cycle of *Saccharomyces cerevisiae*. (Adapted from Hartwell, 1974.)

have condensed chromosomes as the nuclei in more advanced forms of life have.

The total amount of DNA in a haploid *Saccharomyces* cell is about 1.2–1.4 $\times$ 10^{10} daltons. About 80 to 90% of this DNA is contained in the nucleus. Genetic studies have identified seventeen independently segregating groups of centromere-associated genes in diploid *Saccharomyces* during meiosis and sporulation. This number probably represents the complete chromosome complement of the haploid stage of this yeast. There is evidence that other species of yeast have a smaller number of chromosomes. For example, haploid cells of *Hansenula holstii* and *H. wingei* contain three chromosomes, and those of *H. anomala* have two.

VACUOLES

When yeast cells are viewed in the phase contrast microscope, one can usually observe one or more vacuoles of various sizes (0.3–3 μm in diameter). They are especially conspicuous in stationary phase (nonproliferating) cells (Fig. 6). They are generally spherical in appearance and are more transparent to a light beam than is the surrounding cytoplasm. When the cells are placed in nutrient medium and begin budding, the large vacuoles divided by constriction into numerous small vacuoles. During bud development the vacuoles are distributed between mother and daughter cell. And after completion of budding the small vacuoles may coalesce or fuse to form again large vacuoles. Electron microscope observations of thin sections through the cell show that the vacuole is surrounded by a single unit membrane. In freeze-etched specimens both the inner and outer surfaces of the membranes are covered by particles 8–12 nm in diameter; their function is not known, but they may play a role in the transport of substances stored in the vacuole.

Intact vacuoles can be isolated from living cells. One procedure involves converting yeast cells into protoplasts by dissolving the cell wall with lytic enzymes (for example, snail digestive fluid). This treatment is done in the presence of an osmotic stabilizer—for example, 0.8 M mannitol. When the suspension is diluted to 0.4 M mannitol, the protoplasts burst due to swelling by the uptake of water. The vacuoles are more resistant to osmotic lysis. If after a few minutes (when most protoplasts have lysed) the mannitol concentration is restored to 0.8 M, the vacuoles can be isolated by differential centrifugation. The contents of the vacuoles are released when the osmotic stabilizer is diluted with water, and their composition can then be studied.

A great variety of components of both high and low molecular weights can enter the vacuole. Certain cationic dyes (such as neutral red) can penetrate living cells and accumulate in the vacuoles, giving them a pink appearance. This can be demonstrated particularly well in the large vacuoles of the yeast-like organism *Geotrichum candidum*. This phenomenon has been attributed to the presence in the vacuoles of substantial concentrations of negatively charged, polymerized orthophosphate (also termed "volutin" or polymetaphosphate"). The degree of polymerization of this polyphosphate ranges from about 2–10. It is soluble and extractable with dilute cold acid. The vacuoles can also accumulate certain purines or purine derivatives of low solubility (for example, uric acid), causing the formation of conspicuous vacuolar crystals often seen in some species of yeast (for example, *Pichia* species). These bodies are usually in active Brownian movement, for which reason they are sometimes referred to as "dancing bodies." Moreover, a large fraction of the free amino acid pool of yeast is stored in the vacuoles, and some investigators have suggested that lipids are present. When yeast is grown in the presence of excess L–methionine, part of this amino acid is converted in the cytoplasm to S–adenosylmethionine (SAM), an important methyl donor in methylation processes. The excess SAM is carried to the vacuoles by a highly specific transport system and stored in concentrated solution until needed. When such cells are photographed under UV light, the vacuoles appear black due to light absorption by SAM at 265 nm.

The vacuoles also serve as storage vesicles for a number of hydrolytic enzymes, including several proteases, ribonuclease, and esterase. The physical separation of these enzymes from their cytoplasmic substrates prevents cellular breakdown. However, under adverse conditions the vacuoles may break down and autolysis of the cell sets in. Some investigators think that macromolecules and particulate material enter the vacuole by invagination of its membrane, which then surrounds the particle or macromolecule. Eventually the invagination is occluded and the membrane surrounding it breaks down (pinocytosis).

MITOCHONDRIA

Yeast cells have been shown by ultrathin sectioning and by the freeze-fracture technique to contain mitochondria. In *Saccharomyces* they are generally located fairly close to the periphery of the cell. But in *Rhodotorula*, an obligately respiratory yeast, they are more randomly distributed throughout the cytoplasm. In cross sections their diameters range from about 0.3–1 μm, and their lengths from 0.5–3 μm—much, of course, depends on the angle of the section through a mitochondrion. Their number has been claimed to be as high as twenty and as low as one per cell (Fig. 11). During budding the mitochondria become thread-like and may branch, after which they divide and are distributed between mother and daughter cells. The organelles are surrounded by an outer membrane and an inner membrane which forms a number of cristae that extend into the mitochondrial stroma. Mitochondria have been isolated from ruptured vegetative cells by differential centrifugation. They are rich in lipid, phospholipid, and ergosterol, components of the membrane system. They also contain DNA, RNA, and proteins, including RNA polymerase and a number of respiratory enzymes participating in the TCA cycle and electron transport. The prominent function of the mitochondria is that of oxidative energy conversion in the cell.

Mitochondrial DNA (mtDNA) constitutes about 5–20% of the total cellular DNA. The mtDNA from *S. cerevisiae* has a much lower density and (G + C) content (17%) than does the nuclear DNA (about 40% G + C) of this species. The molecular size of mtDNA is about 50×10^6 daltons— much smaller than that of nuclear DNA. There is some evidence that mtDNA is a circular molecule. It codes for a number of respiratory enzymes. Some of the structural proteins of the mitochondria as well as cytochrome C, however, are coded for in the nuclear DNA. The mitochondrial enzymes are translated from mitochondrial messenger RNA into proteins on 70S mitochondrial ribosomes.

Under anaerobic conditions of glucose fermentation or even under aerobic conditions in high glucose concentrations (for example, 5–10%), the mitochondrion appears to degenerate into a form that has been termed a promitochondrion with poorly developed cristae. Such cells no longer sythesize cytochromes aa_3 and b and therefore lack respiratory ability. However, respiration can be restored easily by removing the glucose and aerating the cells in a medium containing a nonfermentable substrate, such as ethanol or glycerol.

A nonreversible loss of respiratory ability also occurs in *S. cerevisiae*. Wild-type or respiratory-sufficient cells (usually designated as ρ^+) continually produce respiratory-deficient mutants in high proportion (1–10%, depending on the strain). Such mutants (ρ^-)can be recognized on malt agar plates by the small colonies (petites) they produce. The cells—which now must depend on fermentation (with its much lower energy production)—therefore produce small colonies. The rate of mutation can be greatly accelerated by growing the cells in the presence of mutagens (for example, ethidium bromide or acridine dyes). This mutation, which has been found to be very stable, involves a deletion of part of, and in some cases all of the mtDNA. Some species of yeast do not produce ρ^- mutants, such as those of the genus *Kluyveromyces*, some haploid *Saccharomyces* (for example, *S. rosei*), and *Schizosaccharomyces pombe*. Such species have been termed "petite-negative yeasts" (See Chapter 5).

MICROBODIES

These organelles, which are especially prominent in certain species of *Candida*, are about 0.5–1 μm in diameter, have a homogeneous matrix, and are surrounded by a single unit membrane. Young cultures of *Candida tropicalis* growing on n–alkanes show numerous microbodies in thin sections, whereas glucose-grown cells show few if any. These bodies resemble "peroxisomes" in higher plants, which contain catalase as a specific enzyme. Enzymatic and cytochemical tests have shown that the microbodies of *C. tropicalis* also have a high catalase content.

ENDOPLASMIC RETICULUM

Ultrathin sections and freeze-etched preparations have shown that the cytoplasm of yeast contains a double-membrane system similar to the endoplasmic reticulum (ER) of higher plant and animal cells. In some preparations the ER appears connected to the outer nuclear membrane, and in others it is found in close association with the plasmalemma. The space between the two membranes of the ER is about 20 nm wide and is filled with an aqueous fluid or enchylema. Sometimes the membranes enclose dense globular bodies up to 100 nm. The cytoplasmic surface of the membranes shows a particulate structure. Some of the clustered particles are thought to consist of polyribosomes, the centers of protein synthesis.

The ER also appears to be involved in bud initiation. ER elements fuse and form a bowl-like envelope around the nucleus and vacuole. This structure then starts to produce a mass of vesicular strands or tubuli, which in turn segregate spherical vesicles that penetrate the plasmalemma. These vesicles probably contain softening enzymes and building blocks for the expanding cell wall.

Some authors have presented evidence that bud initiation is caused by a Golgi-like body or dictyosomes. These Golgi-like structures appear to consist of a stack of three or more flattened sacs or cisternae, surrounded by numerous small vesicles. During cell wall synthesis the Golgi-like apparatus is apparently consumed, but a new Golgi apparatus is formed during subsequent bud formation.

LIPID GLOBULES

Most yeast cells (such as *Saccharomyces* species) contain small amounts of lipid in the form of globules that are stainable with fat stains, such as Sudan black or Sudan red. When used in the proper concentration, these dyes penetrate the cell and accumulate in the lipid globules, staining them bluish-black or red. In electron micrographs of thin sections, the lipid globules show up as the most electron-transparent bodies of the cell, since due to solvent pretreatment they are basically "empty" vesicles. When grown in a medium with a limiting nitrogen supply, some species can accumulate very large amounts of lipid material, sometimes amounting to 50% of the dry weight of the cell. Examples of such yeasts are *Rhodotorula glutinis*, which contains rather numerous lipid globules of various sizes, and *Lipomyces starkeyi* and *Metschnikowia pulcherrima*, which usually have one or two very large lipid globules (Fig. 6b).

THE CYTOPLASMIC GROUND SUBSTANCE

The matrix in which all of the organelles discussed here are located is named the "cytoplasmic ground substance," or more briefly, the "cytoplasm." It contains large quantities of ribosomes, polyphosphates, the storage polysaccharide glycogen, possibly the storage sugar trehalose, a number of glycolytic enzymes, and the little understood 2μm circular DNA.

The ribonucleic acid of yeast is represented by ribosomal RNA, (rRNA), messenger RNAs, and transfer RNAs, the first species amounting to over 85% of the total. Yeast is rich in RNA, containing 50–100 times as much as the DNA. Its concentration depends on the species of yeast, its growth rate, and the composition of the medium of growth. Values for total RNA of 7–12% (dry weight basis) have been reported. Yeast ribosomes have a sedimentation coefficient of 80S, in contrast to the mitochondrial ribosomes, which are 70S particles. The cytoplasmic ribosomes are made up of two subunits, with sedimentation coefficients of 60S and 40S. The 80S ribosomes contain an 18S rRNA species with a molecular weight of about 0.7×10^6 daltons, and a 26S rRNA species whose molecular weight is approximately 1.3×10^6 daltons. The remainder consists of numerous proteins that together with the RNA species form a complex architecture. Most ribosomes in cells are located in polysomes—complexes of messenger-RNA molecules and ribosomes (the number depending on the size of the mRNA molecules). These complexes are actively engaged in polypeptide synthesis.

About 2% of the nuclear genome codes for the rRNA species—a relatively greater percentage than in higher eukaryotic organisms. The rRNA is initially formed in the nucleolar portion of the nucleus as a 38S species, which is subsequently split and processed into the 18S molecule (for the 40S subunit) and the 26S molecule (for the 60S subunit). During processing of the RNA and maturation of the ribosomes, the proteins are attached—also probably in the nucleus. Finally, the 40S and 60S particles assemble to 80S mature ribosomes.

Investigators recently have demonstrated that highly polymerized polyphosphates with degrees of polymerization of 300–500 occur in the cytoplasm in close proximity to the plasmalemma. Since this form of polyphosphate is soluble in cold dilute alkali or in hot perchloric acid, these solvents can be used to extract it from the cell. Polymerized polyphosphates are thought to act as a reserve of high energy phosphate that can be used in various metabolic processes (for example, sugar transport and biosynthesis of cell wall polysaccharides).

Glycogen is one of the two main carbohydrate storage products in yeast. It has a high molecular weight (about 10^7 daltons) and consists of a branched, tree-like molecule. The glucose residues of the main chains are linked by α–(1→4) bonds, and the branches constitute α–(1→6) bonds. There are approximately twelve to fourteen glucose residues between branch points. The glycogen content of yeast varies greatly with the species and the growing conditions. It mainly accumulates in the stationary growth phase when the nitrogen supply is limiting the growth and glucose is still available. Baker's yeast has been reported to contain about 12% glycogen (dry weight basis). Glycogen stains dark brown when the cells are treated with Lugol's iodine. The difference in glycogen storage by various yeast species can be demonstrated by flooding an agar plate with iodine solution after growth of the yeasts for one to two weeks. In ultrathin sections of yeast cells glycogen appears as clumps of spheroidal granular bodies with a diameter of about 40nm. The second carbohydrate storage product is the nonreducing disaccharide trehalose, which may be present from negligible quantities up to 16% of the dry weight basis of the cell depending on the stage of growth. There is some evidence that this sugar is stored in some sort of membrane-bound vesicle to protect it from hydrolysis by the soluble enzyme trehalase (see Chaper 6).

It is difficult to be sure which enzymes are truly soluble in the cytoplasm and which are liberated from cell walls or from a cell organelle during cell extraction or cell rupture. Gentle lysis of yeast protoplasts followed by centrifugation offers perhaps the best guarantee of isolating cytoplasmic enzymes. This technique, however, has not been used by many investigators. It is generally held that hydrolytic enzymes [such as trehalase, maltase, cellobiase, and lactase (the last three enzymes are only in yeast that can utilize the corresponding sugars)], alcohol dehydrogenase, and enzymes of the glycolytic pathway as well as of the oxidative pentose phosphate cycle are located in the cytoplasm.

About 1–5% of the yeast DNA may be present in the cytoplasm, possible associated with a membrane-containing component. This is the so-called 2μm circular DNA, with a density similar to that of the main nuclear component. The molecular weight of this component is about 5×10^6 daltons. Its function in the cell is unknown at this time, although the physical structure of the 2μm DNA is well understood from studies with restriction enzymes.

4 / Sporulation and Life Cycles

In mycology "spore" is a rather general term that indicates a reproductive cell. Both asexual and sexual spores are formed by yeasts. Examples of asexual spores, which are no more than unicellular structures produced by a larger vegetative body or thallus, are blastospores (budding cells formed on a pseudomycelium, Figs. 2e and 3), arthrospores (single-celled fission products of mycelial hyphae, Fig. 2d), endospores (asexually produced within a vegetative mycelial cell), and chlamydospores (rather thick-walled resting spores produced by certain yeasts, Fig 13a). Blastospores (Fig. 2g) are produced by species of many yeast genera (for example, *Candida*) whereas arthrospores (Fig. 2d, 2h) are formed in relatively few genera (for example, *Trichosporon* and *Endomyces*). Endospores are formed only in the genus *Oosporidium* and by some species of *Trichosporon*. Besides serving an

a

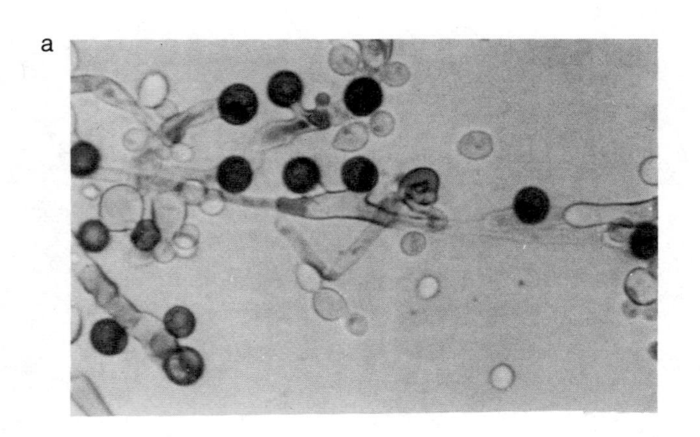

b

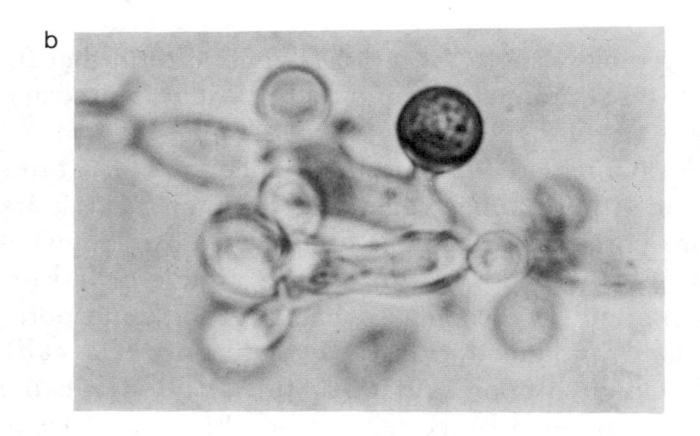

c

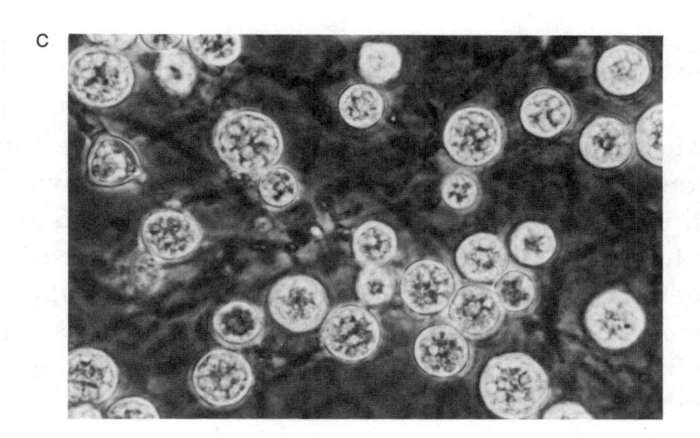

d

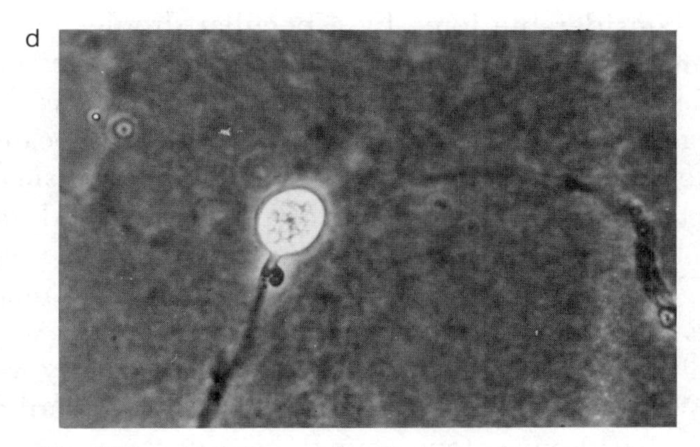

FIG. 13. Chlamydospore and teliospore formation by certain yeasts: (*a*) chlamydospores (dark spheroidal structures) formed *in situ* by *Candida albicans* on trypan blue agar, (note that budding cells and blastospores do not accumulate dye from the medium); (*b*) close-up view of laterally formed chlamydospore; (*c*) teliospores formed *in situ* by *Sporidiobolus johnsonii* (phase microscopy); (*d*) teliospore showing the characteristic clamp connection at the base of the spore.

asexual function—that is, survival under adverse conditions—chlamydospores are occasionally known to be the site of karyogamy and meiosis (sexual) and give rise to asci (*Metschnikowia*) or promycelia in basidiomycetous yeasts (*Aessosporon, Rhodosporidium, and Leucosporidium*). In the latter, chlamydospores that are involved in the sexual cycle are termed "teliospores" (Fig. 13b).

Another type of asexual spore is the ballistospore, which is produced by members of the family *Sporobolomycetaceae* (*Deuteromycotina, Blastomycetes*). These spores are borne, one at a time, on pointed stalks (sterigmata), from which they are discharged with considerable force by a peculiar droplet mechanism. A small droplet of fluid is exuded by the stalk at the base of the spore, and then the droplet is discharged, carrying the spore with it (Fig. 14). Because of the similarity of this process with the mechanism by which sexual spores (basidiospores) are formed and discharged in the higher *Basidiomycotina* (for example, a mushroom), investigators assumed for many years that the yeasts belonging to this group were primitive forms of the *Basidiomycotina*. Recently, van der Walt confirmed this hypothesis when he described a new basidiomycetous genus *Aessosporon* upon the discovery of a sexual cycle in a strain of *Sporobolomyces salmonicolor*, a species producing asymmetric ballistospores. A second genus of the Sporobolomycetaceae, *Bullera*, produces symmetrical (ovoid) ballistospores. The sexual cycle of one *Bullera* species has

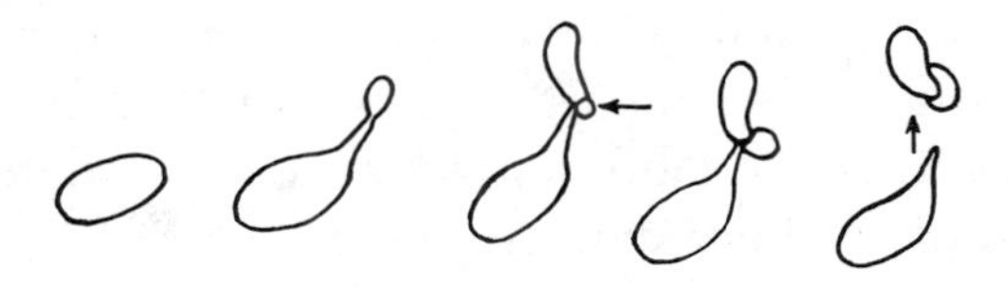

FIG. 14. Formation and discharge of a ballistospore by a species of *Sporobolomyces*.

been observed. The perfect form was also placed in the genus *Aessosporon*.

The formation of sexual spores in yeasts is of great biological significance. Because sporulation constitutes a phase of the sexual cycle of a yeast—that is, alternation of the haploid condition (1*n* chromosomes) and diplophase (2*n* chromosomes). This cycle enables a yeast to undergo genetic recombination, mutation, hybridization, and selection, processes that lead to evolutionary changes.

An essential aspect of sporulation is the reduction division, or meiosis, that a diploid nucleus undergoes. Although the behavior of chromosomes during meiosis is extremely variable in different organisms, the principle of the process is basically similar in all eukaryotic organisms. The nucleus of a diploid yeast cell (for example, *Saccharomyces cerevisiae*) arises through the fusion of the nuclei of two haploid cells and thus contains the 2*n* number of chromosomes. In meiosis the diploid chromosome complement is reduced again to the haploid 1*n* number. A diploid chromosome complement consists of a number (*n*) of pairs of homologous chromosomes. Although yeast chromosomes are too small to be seen under a microscope, the two members of each pair are assumed to have a similar appearance and homologous genetic material on the basis of analogy with higher forms of life. Consequently, the two members of each pair are called "homologous" chromosomes; they are derived from each of the originally haploid cells that fused. During meiosis the homologous chromosomes approach each other and become tightly paired. Each chromosome of the pair consists of two sister chromatids produced during a previous DNA replication. Thus the paired homologues form a four-strand structure, or tetrad. The tetrads (which meanwhile become shorter and more compact) then break apart into two dyads (the two original homologues) that separate during the first meiotic division. Next, the two sister chromatids divide into individual groups of chromosomes (by division of their centromeres). This separation constitutes the second meiotic division. This process results in the formation of four haploid nuclei, each carrying one chromatid of the original tetrad. The haploid nuclei formed in this way are not necessarily identical to those of the two cells that formed the original diploid cell because of the genetic recombination that occurs by the random assortment of the different chromosomes. Also, during the formation of the tetrad, breakage of chromatids may occur, which may be followed by an exchange of parts of the chromatids (recombination) and the rejoining of the broken parts. This interchange of parts between homologous chromosomes, which is termed "crossing over," has been demonstrated in yeasts by genetic experiments (see Chaper 5).

In some yeasts, the four nuclei resulting from meiosis may undergo one additional mitotic division as in the case of *Schizosaccharomyces octosporus*, a species that characteristically forms eight spores in each ascus. Occasionally, further divisions of some or all of the haploid nuclei result in more numerous nuclei in each cell undergoing meiosis. The latter is characteristic of several species of the genus *Kluyveromyces* (for example, *Kl. polysporus*).

In *Saccharomyces cerevisiae* the process of spore delimitation in the ascus has been studied in detail by cytological techniques (see Chapter 3). As shown schematically in Fig. 15 the original nuclear membrane of the diploid nucleus remains intact during the two stages of reduction division, resulting in a four-lobed nucleus in which the four sets of chromatids are separated. Around each of the lobes a structure then forms consisting of a double membrane, the prospore or forespore wall. Spore wall material accumulates within this sac-like structure. The prospore wall thickens and finally encloses the lobe completely, causing the original nuclear envelope to break up into four individual envelopes. Part of the cytoplasm of the original cell is incorporated in the spore by the developing spore wall, but some remains free in the ascus and is referred to as "epiplasm." This process has been named "sporulation by free cell formation" in the early literature to distinguish it from spore for-

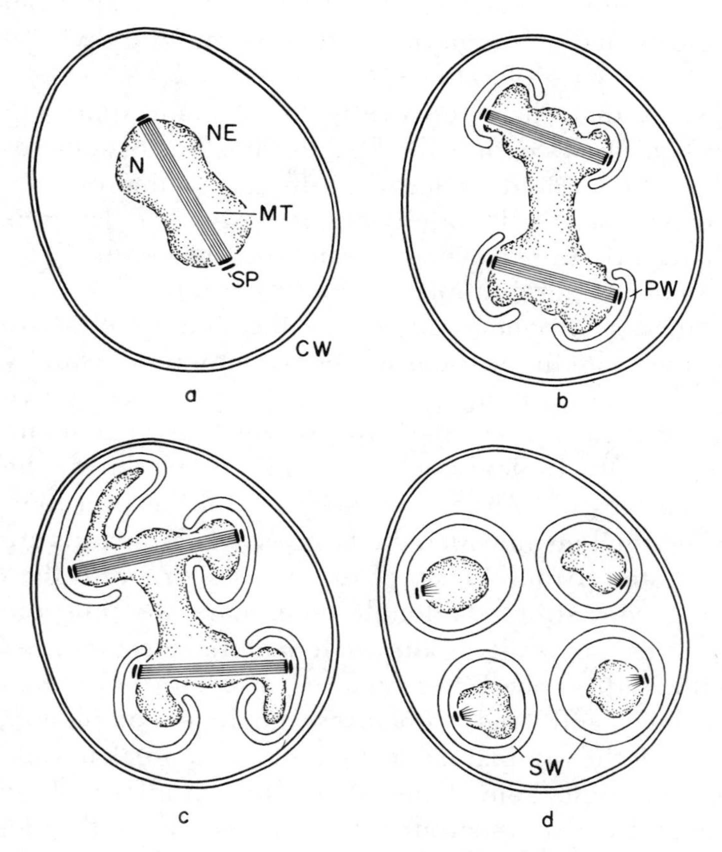

FIG. 15. Diagrammatic representation of nuclear behavior during meiosis and ascospore delimitation in *Saccharomyces cerevisiae* (*CW* =) cell wall; (*N* =) nucleus; (*NM* =) nuclear envelope; (*MT* =) bundle of microtubules; (*PW* =) prospore wall; (*SP* =) spindle plaque; (*SW* =) spore wall. (Adapted from Moens and Rapport, 1971. *J. Cell Biol. 50*, 344.)

mation that is caused by the appearance of cleavage planes through the cytoplasm of the cell (partitioning) characteristic of certain other fungi. Further research is needed to determine if meiosis and spore morphogenesis in other diploid yeasts is similar to that in *Saccharomyces cerevisiae*.

In some yeasts, after sporulation is complete, the ascus wall (the former cell wall) is rapidly digested by endogenous enzymes and the spores are liberated. In other yeasts the ascus wall remains intact, and the spores are liberated only after they swell and rupture the ascus by force. In either event upon germination, the spores can reestablish the diploid condition in a variety of ways, and the cycle can then start over. Many yeasts require special growing conditions to pass through the sexual cycle. If these conditions are not met, a yeast may continue to propagate virtually indefinitely in the vegetative form.

LIFE CYCLES OF ASCOMYCETOUS YEASTS

Ascus-forming yeasts living in various natural habitats can be placed into two broad groupings: those in which the vegetative cells are primarily haploid, and those in which the diplophase predominates. In a diploid yeast (such as *Saccharomyces cerevisiae*) reduction division can occur under suitable conditions directly from the vegetative cells. In haploid cells (for example, all of the species of *Schizosaccharomyces*) the nucleus of the vegetative cell contains the basic ($1n$) number of chromosomes, so that two such nuclei must first fuse to form a diploid nucleus before sporulation can take place.

Although yeasts are commonly referred to as haploid or diploid species, a strict separation into one or the other ploidy is not always possible. In certain yeasts (at least in laboratory cultures), haploid and diploid cells may occur side by side. Hence, in a sporulating culture we may find some asci that were the result of conjugation between two cells and others that arose from diploid cells. On the basis of his extensive studies in the genus *Hansenula*, Wickerham has stated that the most primitive species are exclusively haploid, and the most advanced species are diploid. In between we find species in which various ratios of haploid-to-diploid cells occur. Diploid cells can arise in a haploid culture by conjugation and zygote formation. In some species the zygote then may bud for a number of generations as diploid cells before sporulation takes place.

In diploid yeasts, which produce ascospores directly in the vegetative cells, four mechanisms are known by which the diploid condition is reestablished in the vegetative phase, thus completing the life cycle. These mechanisms (see Fig. 16) are as follows:

1. Two ascospores may fuse directly in the ascus. This is followed by nuclear fusion (karyogamy) of the two haploid nuclei. The first vegetative cell resulting from this zygote (conjugated spores) is a diploid cell. This process occurs in *Saccharomycodes ludwigii* and in some strains of *Saccharomyces cerevisiae*.

2. The ascospores may germinate into relatively small, haploid vegetative cells. After a limited number of generations, fusion occurs between two of these

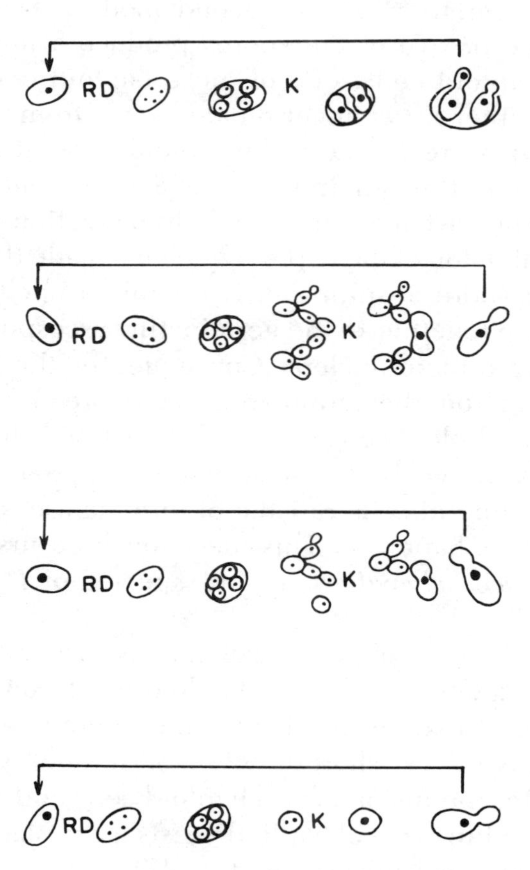

FIG. 16. Four types of life cycles of diploid yeasts (see text for details). Heavy dots represent diploid nuclei; light dots are haploid nuclei; *K*, karyogamy; *RD*, reduction division.

cells, resulting in a diploid generation composed of cells of somewhat larger dimensions than the initial haploid cells. This takes place in some strains of *Saccharomyces cerevisiae*.

3. A variation of the second mode occurs when only one or two of the spores produce a number of small haploid cells; one of these haploid cells fuses with another as yet ungerminated spore from the same ascus, again resulting in the diploid generation. This also is found in certain strains of *S. cerevisiae*.

4. The last mechanism of diploidization consists in the division of the haploid nucleus inside the swelling ascospore, resulting in two haploid nuclei. They then fuse together in the germinating ascospores, producing a diploid nucleus. Consequently, the first bud resulting from the germinating ascospore already constitutes a diploid cell. This mode of action is limited to homothallic yeasts (those in which single ascospores lead to diploid cells capable of forming ascospores in turn; see Chapter 5). This mechanism occurs in *Saccharomyces chevalieri* and in species of *Hanseniaspora*.

The life cycle of yeasts that characteristically spend their vegetative life in the haploid condition is quite different. In yeasts of this type the diploid generation is usually of very short duration, existing only as a zygote after the fusion of two haploid cells and their nuclei. The life cycle of haploid yeasts takes place by the following mechanisms (see Fig. 17):

1. The vegetative cells form tube-like outgrowths (conjugation tubes). The tips of two conjugation tubes

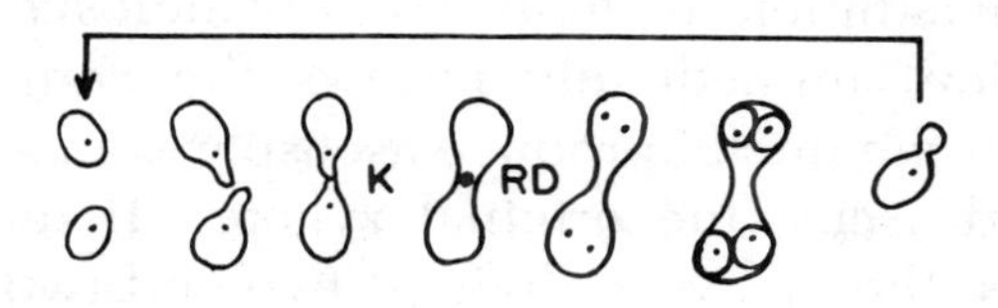

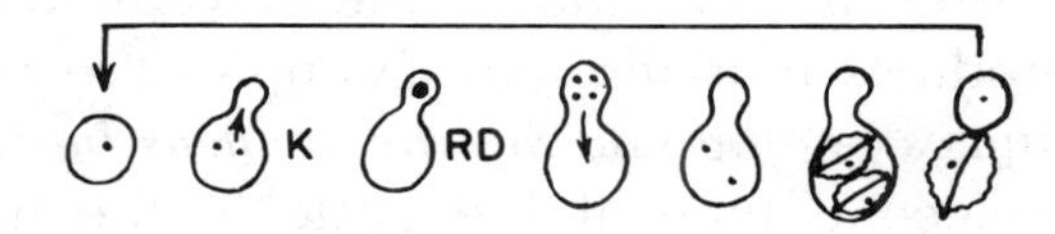

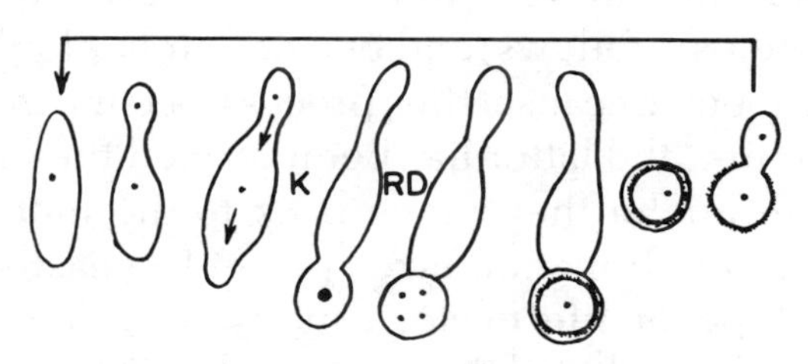

FIG. 17. Three types of life cycles of haploid yeasts (see text for details). Heavy dots represent diploid nuclei; light dots are haploid nuclei; *K*, karyogamy; *RD*, reduction division.

then grow together and a fusion canal is established through which the nuclei of the two cells can approach each other. They then undergo karyogamy (nuclear fusion) either in the fusion canal or in one of the cells, resulting in a diploid nucleus. Usually meiosis (reduction division) immediately follows the conjugation process and the nuclei produce ascospores in a dumbbell-shaped ascus (the original zygote). If there are four spores, they may be divided two and two in the two halves of the dumbbell-shaped zygote, or they may occur as three in one half and one in the other half, and occasionally even as four and zero (Fig. 18). This depends on where the diploid nucleus undergoes reduction division and on the movement of the haploid products thus formed. Examples are *Schizosaccharomyces pombe* and *Saccharomyces bailii*.

2. The second type of life cycle in the haploid yeasts occurs as follows. The haploid cell produces a bud, but the bud does not separate from the mother cell and retains a connective opening of rather large diameter. Nuclear division occurs (mitosis), and the two nuclei then move into the bud where karyogamy occurs. Meiosis follows, and two or four haploid nuclei are produced. Because this process occurs in a budlike structure, the latter has been termed the "meiosis bud." The nuclei then move back to the mother cell and ascospore formation takes place. In yeasts following this type of life cycle, usually only one or two spores are formed, and the extra nuclei, if present, presumably either degenerate or become incorporated in the ascospores. This type of life cycle has been dem-

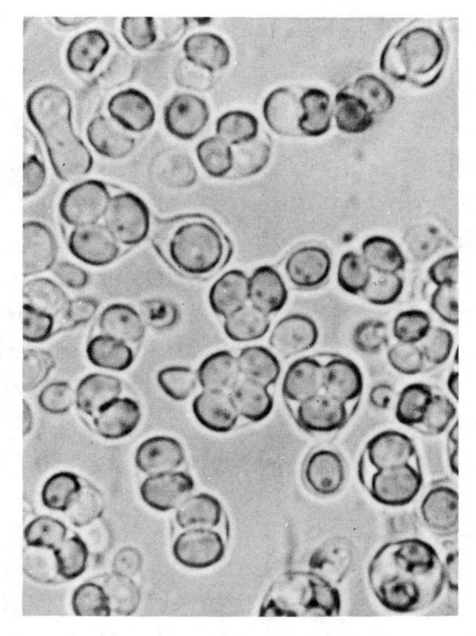

FIG. 18. Asci and ascospores of a hpaloid species of *Saccharomyces* (note dumbbell-shaped asci and spore distribution).

onstrated in the genus *Schwanniomyces*. Recent observations by electron microscopy have revealed that in ascus formation of *Schwanniomyces* and of a *Debaryomyces* species (in which the process of ascospore formation strongly resembles that of *Schwanniomyces*), the bud is first separated from the mother cell by a septum that subsequently dissolves, allowing the two cell nuclei to fuse, and the diploid nucleus undergoes meiosis in the mother cell. By morphological analogy based on the ascus one may presume that one or the other of these two methods is operative in some species of *Hansenula*, *Pichia*, and those species of *Saccharomyces* previously included in *Torulaspora* or Group III *Saccharomyces* according to van der Walt.

3. Another type of life cycle also involves the fusion of two "gametes," but the dikaryotic cell thus formed does not become the ascus. Instead, the nuclei (usually in close association with each other) now move to a specialized cell originating from the dikaryotic structure, which is to become the ascus. In the young ascus—or possibly shortly before they enter—the nuclei undergo fusion, followed by reduction division and ascospore formation (Fig. 17, bottom). In this type of life cycle the gametes may be of different size and shape (*Nadsonia*) or similar in size as in *Eremascus albus*.

Depending on the strain, various species in both the asco- and basidiomycetous yeasts exhibit homothallic (self-fertile) as well as heterothallic sexualities. In the ascomycetous yeasts, as far as is known, species requiring mating types for sexual conjugation are limited to the biallelic, bipolar compatibility system. In this system the determinant for sex is located at one locus on a chromosome. Thus, two compatible alleles result in two mating types termed *a* and *α*.

LIFE CYCLES OF BASIDIOMYCETOUS YEASTS

Since Banno demonstrated in 1967 a basidiomycetous life cycle in strains of *Rhodotorula*, several basidiomycetous genera have been described with varying life cycles.

After fusion of two yeast cells that serve as gametes (1*n* chromosomes) in basidiomycetous genera, a dikaryotic phase (1*n* + 1*n*) usually develops in which the two haploid nuclei do not fuse immediately, but are vegetatively propagated from cell to cell via the clamp connection mechanism (Fig. 19). This mechanism assures that the two compatible nuclei are duplicated in such a way that successive cells contain similar pairs of nuclei. Ultimately, thick-walled teliospores or thin-walled basidia are formed in which the nuclei undergo karyogamy (2*n*), followed by meiosis (reduction division). The resulting four haploid (1*n*) nuclei then arrange themselves either in a promycelium (septate or nonseptate), which develops from the teliospore, or in a basidium. Finally, the basidiospores or sporidia (sexual spores) arise externally. In the basidiomycetous yeasts the sexual spores are not forcibly discharged from the promycelia or basidia; instead, they are formed and reproduce by budding.

The life cycle illustrated in Fig. 20 is representative

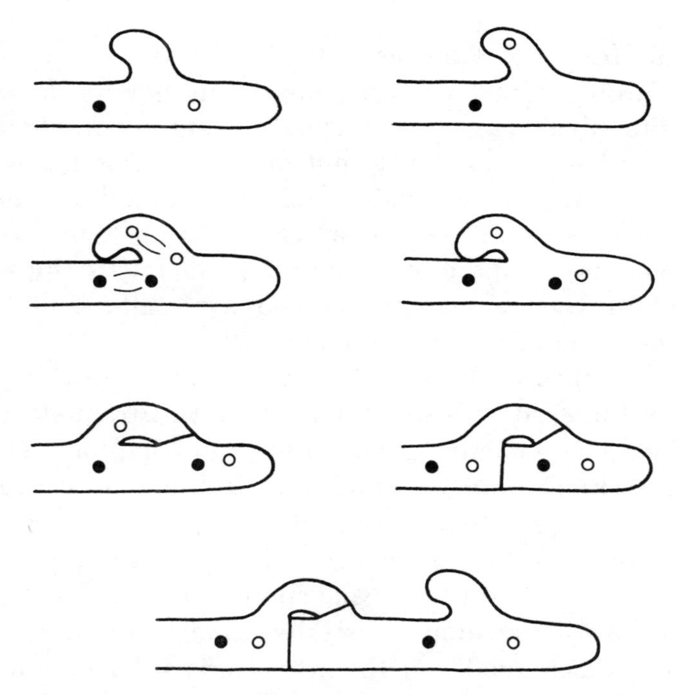

FIG. 19. Schematic representation of nuclear division and clamp formation in a dikaryotic hypha of basidiomycetous yeasts.

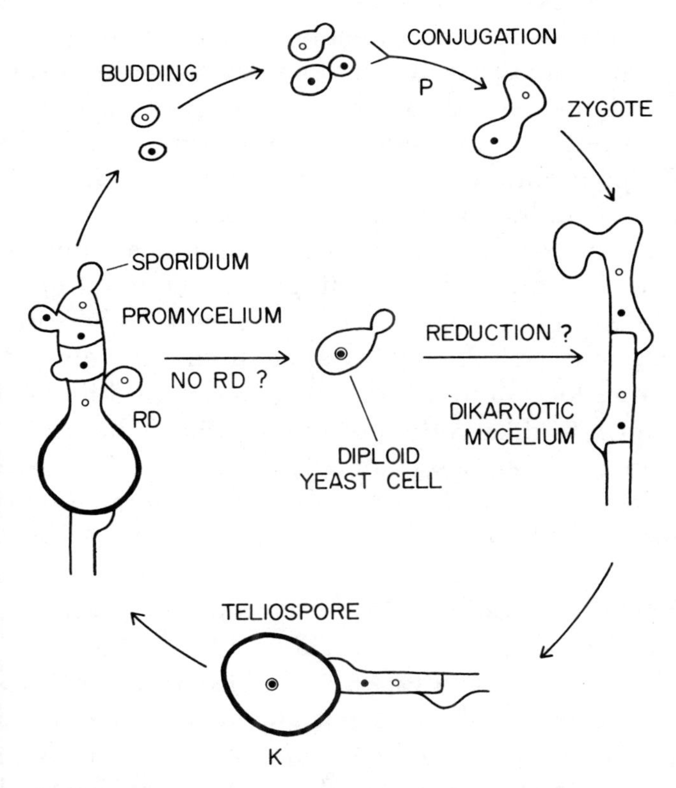

FIG. 20. Diagrammatic life cycle of a heterothallic species of *Rhodosporidium*: *K*, karyogamy; *P*, plasmogamy; *RD*, reduction division.

of those yeasts that form teliospores. Haploid yeast cells arise from the sporidia by budding and propagate in this manner. When cells of compatible mating types fuse (plasmogamy), karyogamy is delayed; instead, a dikaryotic mycelium with clamp connections develops. When conditions are right for sporulation, relatively thick-walled teliospores are formed. The teliospore stage is presumably more resistant, permitting the yeast to survive under adverse conditions. Fusion of the compatible haploid nuclei (karyogamy) occurs either in the teliospore or when the spore germinates to form a promycelium. Normally this is followed by reduction division of the nucleus. The tetrad then supplies the nuclei for the sporidia arising from the promycelium by budding, thus completing the cycle. A diploid phase (as budding cells or mycelium) exists and is believed to arise from the teliospore when reduction division fails to occur. Budding cells produce a dikaryotic mycelium when the diploid nucleus undergoes "somatic reduction," creating a dikaryon. Teliospore-forming genera include *Rhodosporidium*, *Leucosporidium*, and *Sporidiobolus*.

Basidium-forming yeasts, as represented by the genera *Filobasidium* and *Filobasidiella*, have a life cycle as depicted in Fig. 21. The occurrence of a diploid vegetative phase is not known in either genus. The single species of *Filobasidiella* has basidia with a more bulbous apex and produces chains of basipetally formed sporidia, which distinguishes it from the thin basidium with five to nine sporidia apically produced by *Filobasidium*.

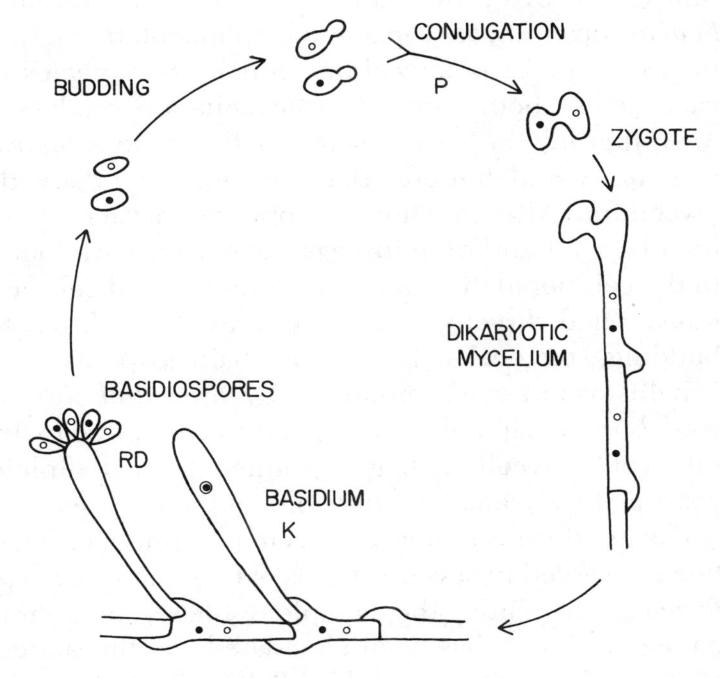

FIG. 21. Diagrammatic life cycle of a heterothallic species of *Filobasidium*: *K*, karyogamy; *P*, plasmogamy; *RD*, reduction division.

In the past decade investigations have shown a variety of ways by which the basidiomycetous yeasts go through their life cycles. Fig. 22 shows variations exhibited by different genera and/or species. Teliospore morphology differs sufficiently to be a useful taxonomic character for species differentiation. The promycelium may be septate or nonseptate, a character used in separating the respective families *Ustilaginaceae* and *Tilletiaceae* in the order *Ustilaginales* (where the basidiomycetous yeasts are placed). Currently, however, two genera, *Rhodosporidum* and *Leucosporidium*, are separated by pigmentation rather than by type of promycelium. Among the species of each genus, both types of promycelia are produced. *Aessosporon*, the sexual stage of the genera *Sporobolomyces* and *Bullera*, does not form a dikaryotic mycelium. After mixing appropriate mating types, both haploid and diploid vegetative stages are found in the cell population. At some point, diploid cells are transformed directly into teliospores. In addition to budding, some cells also produce ballistospores.

Ballistospores are produced by *Sporidiobolus* as well, but the haploid stage appears to be limited to the dikaryotic mycelium that is formed from a diploid yeast cell by "somatic reduction" of the nucleus.

Dikaryotic mycelium with clamp connections has been observed in several strains of *Cryptococcus neoformans*. Recently, the complete life cycle of this pathogenic yeast has been elucidated and the perfect stage has been named *Filobasidiella* (Fig. 21).

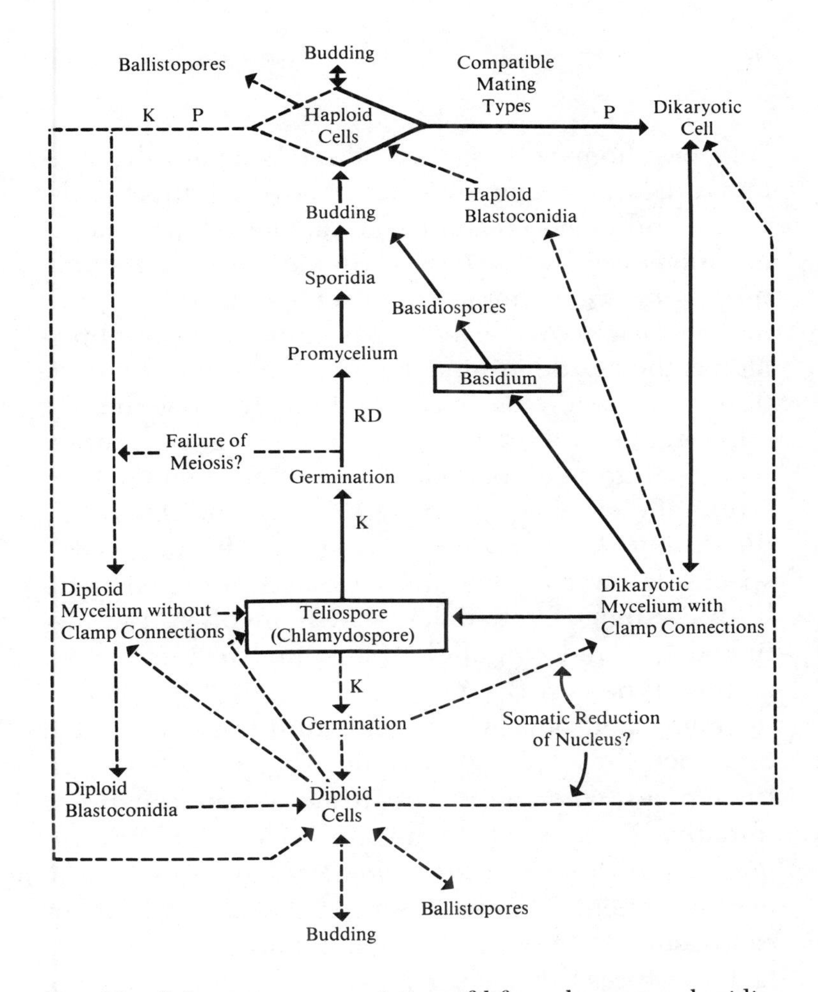

FIG. 22. Schematic representation of life cycles among basidiomycetous yeasts. The life cycle involving the formation of basidia and sessile basidiospores is restricted to the genera *Filobasidium* and *Filobasidiella*. Teliospores, promycelia, and sporidia are found among species of the genera *Rhodosporidium, Leucosporidium, Aessosporon,* and *Sporidiobolus*. *K*, karyogamy; *P*, plasmogamy; *RD*, reduction division; *solid lines,* typical life cycles; *broken lines*, variations.

In basidiomycetous yeasts three compatibility systems are known. Some species have a biallelic, bipolar compatibility system similar to that found in the ascomycetous yeasts. *Rhodosporidium diobovatum* and *R. toruloides* are examples of species with this type of mating system. A species that is multiallelic and bipolar has the sex determinant at one locus, but there are three or more alleles. Such a system gives rise to three (or more) mating types, A_1, A_2, and A_3. Fertile matings will result from all paired combinations involving different alleles as $A_1 \times A_2$, $A_1 \times A_3$, and $A_2 \times A_3$. *Rhodosporidium infirmo-miniatus* is the only yeast species known to have this system. Another compatibility system is tetrapolar, which involves two unlinked loci and two allelic pairs, and results in four mating types: A_1B_1, A_1B_2, A_2B_1, and A_2B_2. Fertile matings occur only when compatibility is satisfied at both loci; thus $A_1B_1 \times A_2B_2$ or $A_1B_2 \times A_2B_1$ are fertile crosses, whereas other pairings are not capable of initiating or completing the life cycle. *Rhodosporidium bisporidiis* and *R. dacryoidum* have this type of mating system. Matings between incompatible types can occur, but growth is usually abortive or lethality factors express themselves and sporulation does not occur. When crosses are made in tetrapolar systems with incompatible mating types where only one locus is satisfied (for example, $A_1B_1 \times A_2B_1$), conjugation has been observed, but the mycelium is poorly formed and does not develop clamp connections or teliospores.

INDUCTION OF SPORULATION

One of the first tests used in identifying a yeast after it has been obtained in pure culture is to determine its ability to form asco- or basidiospores. However, the life cycle of a yeast does not consist of a rhythmic alternation of generations. Some yeasts will continue to grow vegetatively for many generations, and only under very special conditions can the sexual cycle be induced. In heterothallic, haploid yeasts mixing of compatible haploid strains is necessary before sporulation can occur. In the following section, therefore, we consider the questions "How does a yeast sporulate?" and "What induces a yeast to form sexual spores?"

At one time investigators believed that a yeast would sporulate only under conditions unfavorable for growth (for example, yeast placed on blocks of gypsum wetted with distilled water). Today, however, investigators generally accept that sporulation in yeasts is not necessarily the result of the culture's exposure to unfavorable conditions. Evidence of this is found in the heavy sporulation of isolates of many yeasts, which have recently been obtained from natural sources, on rich nutrient media such as malt agar, yeast autolysate-glucose agar, and other commonly used isolation media. Young colonies two- to three-days old may develop on such media masses of asci and ascospores, or more rarely teliospores. In most instances yeasts cultured in the laboratory are propagated on richer media than were available to them in their natural habitat from which they are isolated. Conversely, many cultivated yeasts—particularly baker's and brewer's yeasts, as well as yeasts isolated from spoiled food or found as contaminants in industrial fermentation processes—sporulate only with difficulty on media rich in nutrients. Special media have to be employed to induce a high percentage of the cells of these organisms to sporulate. We believe that one reason for finding so many strains in a natural environment that sporulate heavily on malt agar is that in such habitats ascospores survive better than vegetative cells. This greater resistance is probably due to the thick wall of the spores. Ascospores are more resistant to freezing, drying, exposure to harmful chemicals, and exposure to high temperatures. But, in contrast to the very heat-resistant bacterial endospores, the heat resistance of ascospores is only 6–12°C higher than that of vegetative cells under comparable conditions. One should actually state that the death rate of a sporulated culture of yeast is lower than that of a vegetative culture. For example, Wickerham reported that four minutes at 58°C were sufficient to kill a vegetative culture of *Hansenula anomala*, but a sporulated culture still contained viable spores after ten minutes, although none were left after longer heating. Similarly, at the turn of the century Beijerinck demonstrated that yeasts that had been kept in culture collections for a long time and that tended to lose the ability to sporulate could be reconverted into more heavily sporulating cultures. He did this by treating them with toxic agents, such as high concentrations of ethanol, that preferentially destroyed the vegetative cells as compared with the spores. That some yeasts tend to lose their ability to form ascospores in culture over a period of years is well known and may possibly be ascribed to more rapid growth of permanently asporogenous vegetative forms on laboratory media as compared with cells entering the meiotic cycle.

Numerous sporulation media and techniques have been described and advocated in the literature for the purpose of increasing the percentage of sporulating cells in comparison with the normal cultivation media. Some authors have recommended first growing a yeast on a complex and rich presporulation medium so that sporulation on special media can be facilitated. A yeast culture should be freshly transferred and well nourished for effective sporulation to occur. In a study of sporulation in *Saccharomyces*, investigators found that young daughter cells sporulate poorly or not at all, but they sporulate normally after producing a daughter cell in turn. Complex sporulation media (for example, a mixture of vegetable juices with or without autoclaved baker's yeast) have been devised for the purpose of satisfying the sporulation requirements for as many yeasts as possible. However, no sporulation medium can be called "universal." The compositions of several special sporulation media are shown in Table 1 (distilled water is used in all media preparations).

Of these, acetate agar is particularly effective for species of *Saccharomyces*, Gorodkowa agar for species of *Debaryomyces*, potato-glucose agar for certain hap-

TABLE 1. The compositions of several special sporulation media (in percent)

Ingredients	Yeast autolysate-glucose agar	Gorodkowa agar	YM agar (after Wickerham)	Corn meal agar	Acetate agar (after Kleyn)	Potato-glucose agar
Yeast extract or autolysate (dehydrated)	0.5		0.3			
Glucose	5.0	0.25	1.0		0.062	2.0
Agar	2.0	2.0	2.0	2.0	2.0	2.0
Meat extract		1.0				
NaCl		0.5			0.062	
Malt extract			0.3			
Peptone			0.5			
Corn meal extract[a]				(100 ml)		
Bacto-Tryptose					0.25	
Na-acetate. $3H_2O$					0.5	
Potato extract[b]						(23 ml)
Distilled water						(77 ml)

[a] 12.5 g of yellow corn meal is stirred into 300 ml of water and heated in a water bath at 60°C for one hour. The suspension is then filtered through paper. The volume of the filtrate is made up to 300 ml and autoclaved. Corn meal agar is also available commercially.

[b] Potato extract is prepared by grinding 100 g of whole potato tissue. Add 300 ml of water and store overnight in a refrigerator. Filter through several layers of cheesecloth and autoclave the filtrate. Potato-glucose agar is also available commercially.

loid species of *Saccharomyces*, yeast autolysate-glucose medium for the production of asci in *Schwanniomyces* and related yeasts with mother–daughter cell conjugation, YM agar for many species of *Hansenula* and *Pichia*, and corn meal agar for several yeast-like fungi (for example, *Endomyces* and *Saccharomycopsis*).

It is generally agreed that aerobic conditions are essential for sporulation, since few, if any spores are formed under anaerobic (fermentative) conditions. Most work on the physiology of sporulation has been done with strains of *Saccharomyces cerevisiae*. For maximal sporulation, the cells should be preadapted to respiratory growth. For example, cells harvested during log phase growth on glucose (fermentative metabolism) sporulate poorly on acetate medium; but cells growing in log phase on a substrate that can be metabolized only by respiration sporulate abundantly on acetate medium. However, these findings appear to apply primarily to *Saccharomyces cerevisiae* since there are many species that respond quite differently to environmental conditions. For example, a slightly reduced oxygen tension appears to be very favorable in stimulating spore formation in the species of *Hanseniaspora* that produce asci with a single spheroidal spore. When these species are grown on potato-glucose agar and part of the inoculum is covered with a coverslip, large numbers of asci usually develop under the coverslip in areas very close to the edge, but few develop in the aerobic, uncovered portion of the growth (Fig. 23a).

a

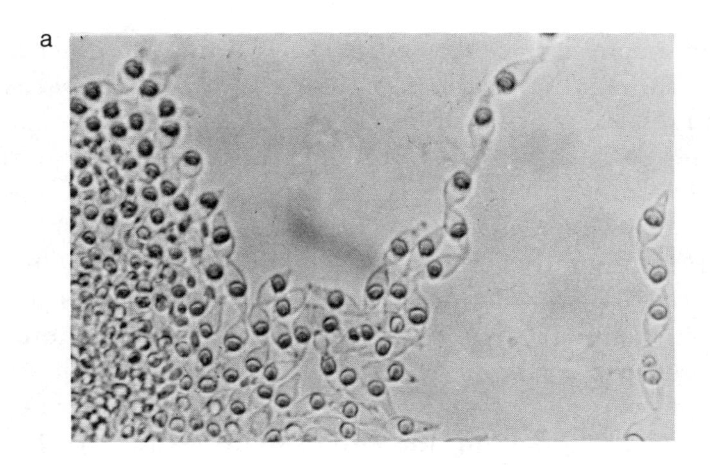

b

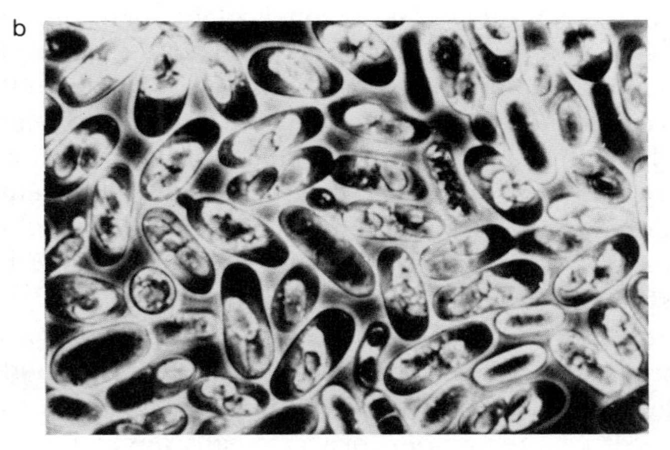

FIG. 23. Ascospore formation by certain yeasts influenced by environmental conditions: (*a*) large numbers of asci of *Hanseniaspora uvarum* formed under reduced oxygen tension; (*b*) asci of *Cyniclomyces guttulatus* formed at temperatures nonpermissive for vegetative growth. See text for details.

The pH of the media is not particularly critical, but an optimal range appears to exist between pH 6 and 7. The temperature seems to be optimal in the normal ranges of room tempterature, approximately 20–25°C. But *Cyniclomyces guttulatus*, which grows only in the range of 30–40°C, can sporulate only at nonpermissive temperatures for vegetative growth, the optimum being 17–18°C (Fig. 23b).

The time required for sporulation to occur and the percentage of cells converted to asci vary widely with species, variety, and even strain of yeast. Species freshly isolated from nature sometimes sporulate in one or two days, whereas others require one to two weeks. Strains held under laboratory conditions for an extended period of time may require from one to several weeks before spore formation becomes apparent. For this reason, organisms on sporulation media should be investigated at frequent and regular intervals, for once formed, spores may undergo germination on the same medium.

Many yeasts in which sexual spores could not be observed and therefore classified in imperfect genera (for example, *Candida*) were later found to be haploid, heterothallic strains that readily sporulated when mixed with a compatible mating type. For this reason, when a number of similar asporogenous isolates are obtained from natural sources, the isolates should be mixed on sporulation media and the growth be inspected for zygotes and spores. This is usually done in groups of six or eight. If mating is observed in a group, characterization of mating types can be obtained by further mixing in fours or pairs. If a particular species of asporogenous yeast is known to be a haploid mating type and an unknown isolate keys to that species, mixing of the isolate with such mating types is done to confirm interfertility.

TAXONOMIC VALUE OF SEXUAL SPORULATION

The ability or inability to produce sexual spores is the primary criterion for yeast's placement in one of the three subdivisions of fungi. These sexual spores formed by various yeasts exhibit a wide variety of form, surface markings, size, color, inclusion bodies, and in number per ascus. Fig. 24 depicts a number of shapes of ascospores formed by various yeasts. Generally most of these features are quite constant for a given species. In some genera (for example, *Nadsonia* and *Schwanniomyces*) the respective species have a similar spore morphology, but in other genera (for ex-

FIG. 24. Examples of ascospore shapes in yeast. *Left to right*: spheroidal, ovoidal, kidney or bean shaped, crescent or sickle shaped, hat shaped, helmet shaped, spheroidal with warty surface, walnut shaped, Saturn shaped, spheroidal with spiny surface, arcuate, needle shaped with appendage. Spores of certain species often contain lipid globules of various sizes. See text for examples of species containing these spores.

ample, *Hansenula* and *Pichia*) spore morphology may vary among the species. For example, *Hansenula capsulata* forms hat-shaped spores, whereas those of *H. saturnus* are Saturn shaped. Spherical or short ovoid (globose) spores with a smooth surface are commonly found in *Saccharomyces* (Fig. 25a), with the exception of those which show mother-daughter cell conjugation and which have a warty spore surface. This feature, which can be seen only in the scanning electron microscope, is found among the species that van der Walt has placed in *Saccharomyces* Group III, and of which several were placed at one time or other in the genus *Torulaspora*. Although appearing smooth and spheroidal in light microscopy (Fig. 25b), the spores of *Saccharomycodes* have been shown in thin section by TEM to possess a short brim, and thus they are technically "hat-shaped." Spores with the shape of a kidney bean are characteristic of species of *Kluyveromyces* (for example, *K. fragilis*, Fig. 25g) and of some species of *Schizosaccharomyces* (for example, *Schizosaccharomyces japonicus* var. *versatilis*). *Saccharomycopsis* (= *Endomycopsis*) *selenospora* forms lunate or asymmetrically fusiform (crescent-like) spores. Eight long, thin, needle-shaped spores with a nonmotile, whip-like appendage are found in *Nematospora* species (Fig. 25h, 25i), whereas species of *Metschnikowia* produce thin needle-shaped spores without appendages (Fig. 25j). *Metschnikowia* spores are pointed at one or both ends, depending on species. Arcuate spores (Fig. 25c) are formed by *Crebrothecium* (= *Eremothecium*) *ashbyi*.

a

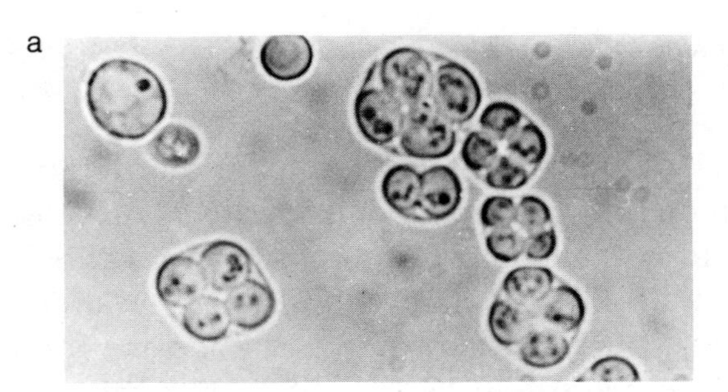

b

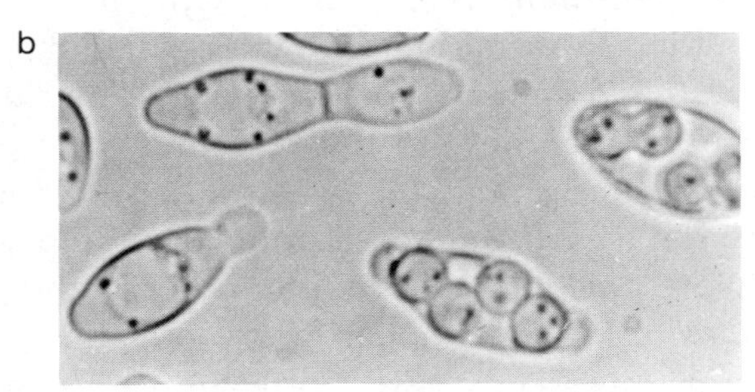

c

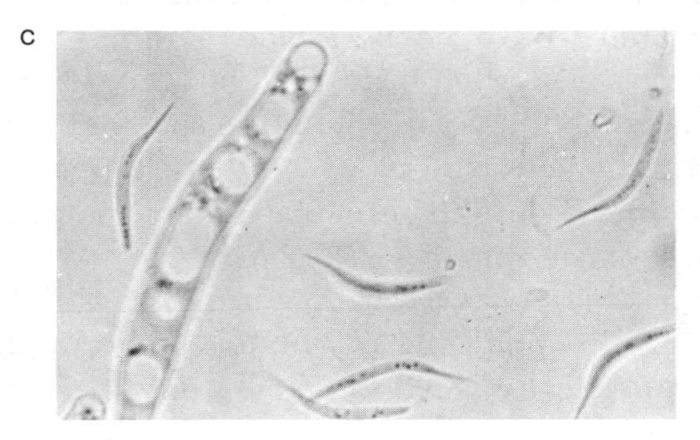

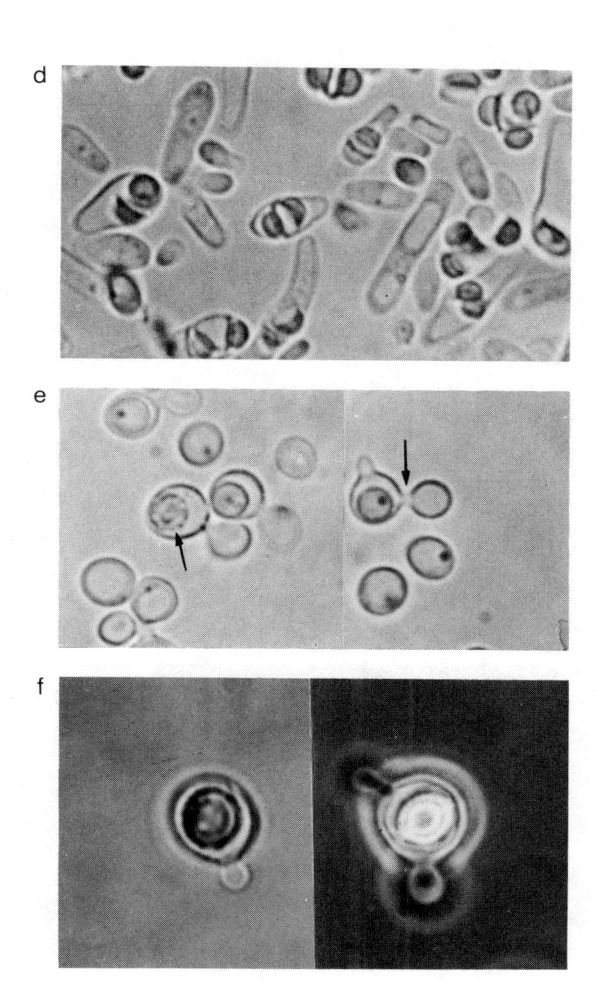

FIG. 25. Asci and ascospores of yeast: (*a*) spheroidal spores, two to four per ascus in a diploid *Saccharomyces* species; (*b*) an ascus with two pairs of spheroidal spores (actually they possess a very short brim) in *Saccharomycodes ludwigii*; in one ascus two spores are conjugating; the refractile granules are presumably polyphosphate storage products; (*c*) arcuate spores of *Eremothecium ashbyi*; (*d*) asci and liberated spores in *Hansenula anomala*; the spore shape resembles a derby hat; (*e*) asci of *Debaryomyces hansenii* with a single (usually), warty, spheroidal spore. Note fusion between two cells or between a cell and a bud; (*f*) asci of a *Schwanniomyces* species with walnut-shaped spores. Note difference in appearance of spore when observed under phase contrast (*right*) and bright-field illumination (*left*).

g

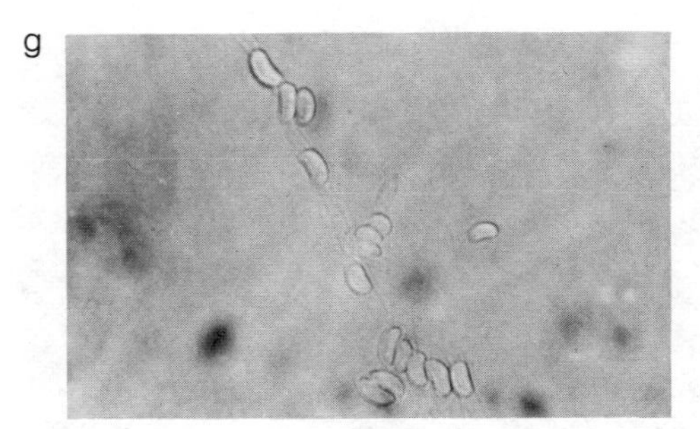

h

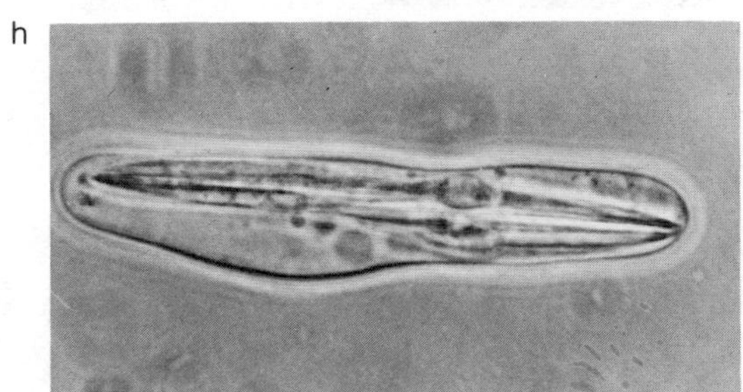

i

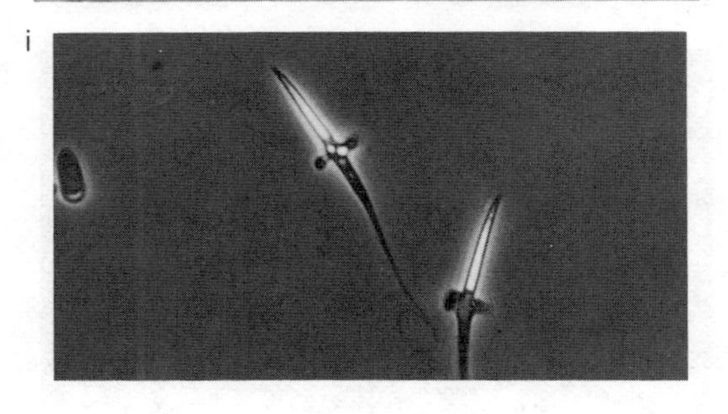

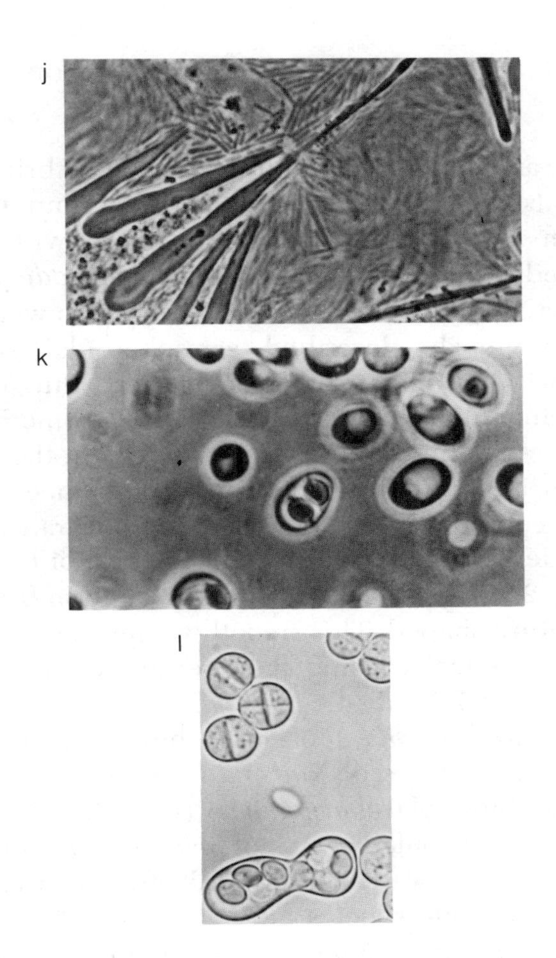

FIG. 25 (*continued*). (*g*) Reniform (kidney-shaped) ascospores formed by *Kluyveromyces fragilis* are liberated from the ascus at maturity; (*h*) needle-shaped ascospores within an ascus of *Nematospora coryli*; (*i*) germinating ascospores of *N. coryli* (note the characteristic nonmotile, whip-like appendage); (*j*) club-shaped asci and needle-shaped ascospores produced by *Metschnikowia bicuspidata* within the body cavity of a brine shrimp (*Artemia salina*); (*k*) Saturn-shaped ascospores in an ascus of *Hansenula saturnus*; (*l*) ovoid spores in a dumbbell-shaped ascus of *Schizosaccharomyces octosporus*.

There are also spores that have a ledge, or brim, that surrounds the spores. The degree of prominence of the brim and its position on the spore vary widely. As explained earlier for spores of *Saccharomycodes*, electron microscopy has demonstrated a brim on what had been previously described as spheroidal spores. Those with a helmet-shape have short, thin brims, nearly tangential to the spore surface (in some *Pichia* species); sometimes the brim is so low that the spore may appear hemispherical. A more pronounced brim, slightly curved, gives the spore the appearance of a typical derby hat as seen in some species of *Hansenula* (Fig. 25d). Another type of spore found in *Hansenula* is Saturn shaped. This smooth-walled spore has an equatorial ledge, causing it to resemble the planet Saturn (Fig. 25k).

The topography of a spore may show other interesting features. Spores of *Nadsonia* are spiny surfaced (Fig. 26a), those of *Debaryomyces* (Figs. 25e, 26b), *Citeromyces* and some species of *Schizosaccharomyces* have irregular or warty surfaces. Electron micrographs of spores of some species of *Pichia* (Fig. 26c) and of *Saccharomyces*, originally thought to be smooth walled, have shown varying degrees of surface irregularity (wartiness). When observed by light microscopy, the spores of *Schwanniomyces* appear convoluted with an equatorial, thickened brim, causing the spore to resemble a walnut (Fig. 25f). However, when observed by scanning electron microscopy, the spore topography is clearly warty rather than convoluted (Fig. 26d). As can be seen in Fig. 26, the size and density of the warts on the spores varies among the different species. Ultrathin sections of spores viewed by TEM, have revealed that the layering of the spore wall differs between *Debaryomyces*, *Pichia*, and *Saccharomyces* species.

In most genera the spores are without color (hyaline). In some genera, however, the spores are amber to brownish in color (*Nadsonia* and *Lipomyces*), which is evident upon observation in the light microscope. Formation of teliospores also gives a brownish caste to the normally reddish colonies of *Rhodosporidium* and *Sporidiobolus*. In some species of other genera, colonies of heavily sporulating strains appear light pink to reddish-brown due to the production of pulcherrimin (See Chapter 6), but the color of individual spores is not sufficiently pronounced to be apparent under the microscope. This pigment is found quite often in species of the genus *Kluyveromyces*, and less fequently in *Pichia* and *Hansenula*.

The number of spores per ascus may vary for a particular species, although the majority of the yeasts form a definite maximum number of spores in each ascus. In some yeasts the maximum number is eight (for example, *Schizosaccharomyces octosporus*, Fig. 25l). But in most ascosporogenous yeasts it is four (*Saccharomyces cerevisiae*, Fig. 25a), although asci with two or three spores in this species are quite common. Occasionally, due to additional mitotic divisions before spore formation, asci are noted with five, six, or more spores. There are also yeasts that produce only one or, more rarely, two spores per ascus (*Schwannio-*

a

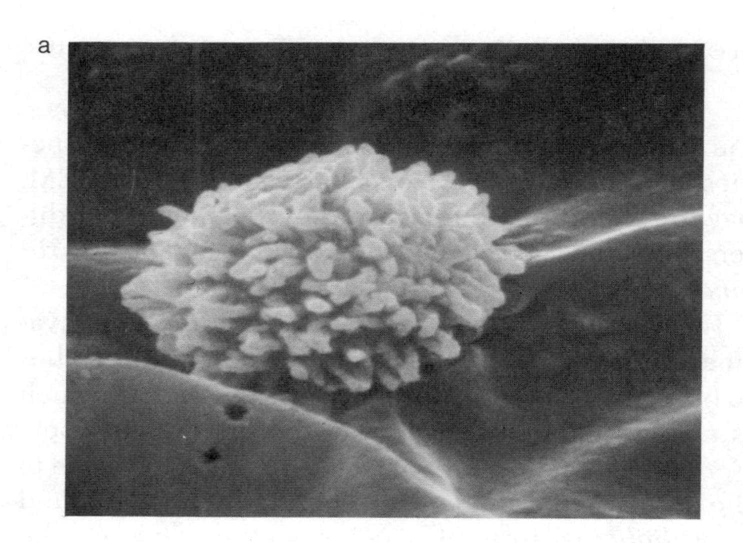

b

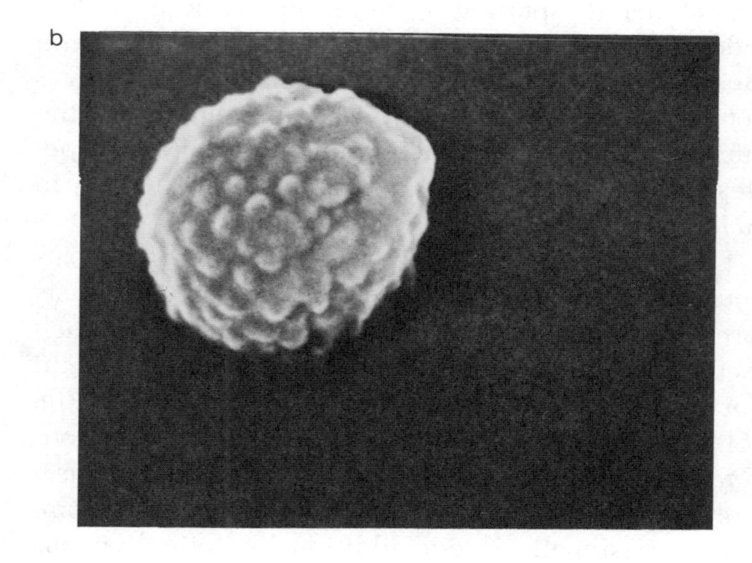

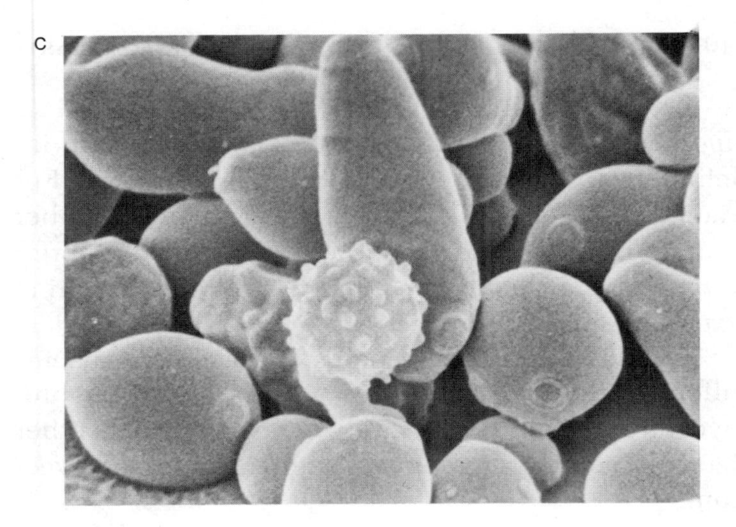

FIG. 26. Scanning electron micrographs of ascospore ornamentation: (*a*) "spiny-surfaced" ascospore of *Nadsonia elongata*; (*b*) warty spore of *Debaryomyces hansenii*; (*c*) sparsely warted surface of a spore of *Pichia polymorpha;* (*d*) warty surface and thickened equatorial ledge of spores of *Schwanniomyces occidentalis*.

myces, *Debaryomyces*, Fig. 25e, 25f). Yeasts with indefinite numbers of spores are also known. For example, *Lipomyces* produces a variable number ranging from four to sixteen and occasionally more. *Kluyveromyces polysporus* has asci often containing sixty-four spores or more.

Size and shape of teliospores are useful in species differentiation of basidiomycetous yeasts, but the promycelia and sporidia are rarely of taxonomic use other than the septate or aseptate nature of the promycelium.

All of these characteristic features, including the manner of ascus or teliospore formation, are of great taxonomic value in separating the yeasts into their respective subdivisions, orders, families, and genera—and in some cases even species within a genus.

We prefer to observe sporulation and spore morphology in wet mounts of living material. Although ascospores can be stained with malachite green and other basic dyes as one would stain bacterial endospores, certain morphological details are lost during the staining process. Furthermore, the ability of yeast spores to retain dyes varies considerably with the species and the age and condition of the spores. Since ascosporulation may be somewhat localized in the cell mass on a slant of sporulation medium, we customarily examine living material from the surface of both the thick (butt) and thin (tip) portions of the slant. Teliospore formation usually occurs in the subsurface portions of the medium, and consequently may be missed if not properly sampled.

HETEROTHALLISM AND SPORULATION

While the occurence of heterothallism in yeast was recognized by Lindegren in 1943, its occurrence in other fungi has been known much longer. It commonly occurs among the various groups of yeast and its importance was noted in relation to the yeasts' life cycles. However, due in part to the classical technique of obtaining pure cultures by selecting single, well-isolated colonies, many species were described as lacking a sexual phase. The recognition of heterothallism has aided in the proper placement of many of these organisms, where the mixing of compatible strains of a species resulted in sporulation.

In the transition from haploid vegetative cell division to sexual conjugation, the former process is arrested and the cells fuse, usually after the formation of conjugation tubes. Conjugation occurs normally among cells in the unbudded portion or G1 phase of the cell cycle—that is, prior to duplication of the nuclear spindle plaque (Fig. 12). It is thought that in heterothallic yeasts the arrest of mitosis (or of DNA replication) and the formation of conjugation tubes is caused, at least in part, by two sex pheromones, α factor and *a* factor, which have been studied extensively in mating types of heterothallic *Saccharomyces cerevisiae* strains. Cells of mating type α excrete a small copper-containing peptide with a molecular weight of about 1,400 daltons. Very recently Tanaka and coworkers elucidated the entire amino acid sequence of the peptide. They could not confirm the presence of copper in the molecule. This peptide, which is produced constitutively by α cells, only affects *a* cells. An *a* factor also has been identified that is produced constitutively by *a* cells, and it affects only α cells. Its chemical composition is not known.

The mating reaction between naturally occurring, heterothallic types is quite variable. Sometimes the reaction is so strong that it is preceded by a pronounced agglutination between cells of compatible mating types, which is then followed by zygote formation. With other species, however, the mating reaction may be very weak and difficult to detect, possibly because incompatibility factors are present in the mating types. In such cases it is very helpful to prepare mutants of the mating types that have a nutritional deficiency (auxotrophs). This can be done by treating the cultures with ultraviolet light or using chemical mutagens. For example, we mix on a sporulation medium one mating type requiring adenine and another requiring L–arginine for growth. After sufficient incubation the mixture can be streaked on a synthetic medium that contains neither adenine nor arginine. Only zygotes will grow on the "minimal" medium, because each partner supplies what the other lacks. In this way one zygote can be detected and isolated among a million or more nonsporulating cells, and sporulation of the diploid zygote (or its offspring) can be conveniently studied.

Sexual differentiation in yeasts does not imply that yeasts have reproductive structures to which the terms male and female gametes can be applied (see Chapter 5). The term "gamete" is normally reserved for highly differentiated male and female fusion cells rather than for the relatively unspecialized vegetative cells, ascospores, or sporidia of yeast. Nevertheless, conjugating yeast cells, ascospores, or sporidia are often designated as gametes—especially if these gametes represent different mating types. Because such gametes are not differentiated morphologically into male and female structures, they are usually designated as *a* and α, A_1, A_2, and A_3 or A_1B_1, A_1B_2, etc. The fusion of haploid, homothallic cells prior to sporulation may be considered as a case of somatic conjugation.

PARASEXUALITY IN YEASTS

Various investigators have discovered situations where yeasts (and other fungi) alternate between the haplophase and diplophase without the formation of sexual spores. Pontecorvo termed this process "parasexuality" when he observed it to occur in *Aspergillus nidulans* in 1954. Subsequently, it has been observed in a number of other fungi, including species of *Metschnikowia*, where the process was termed "protosexuality" when observed by Wickerham. This mode of alternating haplo- and diplophase is operative not only in fungi without known perfect stages, but also in yeasts where a known ascal stage has been established. Van der Walt has likened the diploid structure (cell) to the dangeardien (*sensu* Mareau), which delimits the haplophase externally (by budding). The diplophase is reestablished by conjugation between haploid cells or by autodiploidization of a haploid cell. In known cases, parasexuality occurs relatively infrequently. However, detection of this process is difficult and the overall significance is not known. The simultaneous existence of both haploid and diploid cells could be of advantage to a yeast. In diploid cells mutant recessive genes can spread, masked by the wild type in the heterozygote, thus providing a stable phenotype along with the possibility of genetic variation. The haploid cell, however, can make possible the rapid appearance of mutant characters that would be masked in a diploid. Such mutant genes could have potential advantage in heterozygous diploids. Thus with both phases coexisting in genetically separated states and being able to reproduce asexually indefinitely, the yeast can have the advantage of haploidy and diploidy at the same time.

Since sporulation governs to a large extent the evolutionary development of the various groups in the phylogenetic lines, it is undoubtedly the most important biological event from the standpoint of the yeast itself. To the investigator, yeast sporulation has made possible genetic studies. It also is one of the principal foundations upon which yeast taxonomy is based. In spite of occasional difficulties encountered in the study of sporulation, its importance more than justifies the time invested in trying to establish sporulation conditions for cultures isolated from various sources. Finally, the observation of sporulating yeasts and their asco- or teliospores is indeed a thrilling experience in the study of yeast.

5 / Yeast Genetics

Although the formation of ascospores in yeast was already recognized during the middle of the nineteenth century, sporulation in diploid *Saccharomyces* was thought to be associated with some type of parthenogenetic process, a term found over and over in the older literature. In parthenogenesis (best known in some members of the insects and crustacea) sexual reproduction can occur with only one sex cell (the female egg) involved, the male gamete being absent. It was believed that such a process occurred in diploid yeasts, such as *Saccharomyces cerevisiae*, since there was no evidence that sexual conjugation took place. The true significance of ascospore formation in yeast was discovered independently by Satava in Czechoslovakia and Winge in Denmark during the years 1934–1935. The fundamental discoveries of these investigators established that in *Saccharomyces* a true alternation of generations occurs. This means that the vegetative cells of the yeast are diploid (containing the $2n$ number of chromosomes), whereas the ascospores were found to be haploid and to contain $1n$ chromosomes. By a fusion process (conjugation) between the ascospores or their progeny, the diplophase was reestablished. This discovery of the alternation of haplophase and diplophase in yeast (similar to the life cycles of higher organisms) made productive genetic studies with yeasts possible. Cultures resulting from single ascospores can be used for investigations on genetic segregation, mutation, and hybridization, especially since ascospores of many strains of yeast remain haploid when allowed to germinate individually.

Winge and his collaborators developed techniques for isolating the four ascospores of strains of *Saccharomyces cerevisiae* by using a micromanipulator. By observing and comparing the topography of giant colonies produced when isolated spores were allowed to grow on malt gelatin media, genetic segregation was shown to occur in species of the genus *Saccharomyces.* In these early days of yeast genetics, Winge and his coworkers showed that it is possible to hybridize spores produced by two different species of *Saccharomyces.* They placed a spore from one species and that from another side by side in a droplet of medium. When they grew together (conjugated), they produced a diploid offspring, representing a hybrid. If one of the parents fermented a certain disaccharide and the other did not, the hybrid was always able to ferment this sugar. In other words, the genes responsible for the synthesis of specific hydrolytic enzymes (necessary for the hydrolysis of disaccharides to be fermented) are dominant in the F_1 generation.

Another great step forward was the fundamental discovery by the American investigator Lindegren in 1943 that heterothallism exists in *Saccharomyces.* He discovered that in many strains of *Saccharomyces cerevisiae,* two of the ascospores belong to one mating type and the other two to another mating type, usually referred to as *a* and α. This 2:2 segregation indicated that mating is controlled by a single gene. Lindegren found that if the *a* and α spores were allowed to germinate separately, they developed into haploid clones, or populations. However, when *a* and α populations were mixed, conjugation between the cells of opposite mating types occurred rapidly and the diplophase was reestablished. The finding of haploid clones of opposite mating types has greatly facilitated genetic studies with yeast, because it then became possible to produce various mutants and do many genetic crosses with the offspring of one single ascospore.

Two complications in this technique of hybridization have been encountered. One is the possibility that members of a haploid population may conjugate with cells of their own mating type, forming a homothallic diploid yeast. This process, which occurs in populations of cells of the α mating type, is called "selfing" or "selfdiploidization." Another mechanism by which diploids homozygous for mating type are produced is by endomitosis—that is, chromosome duplication in the absence of nuclear division and budding. Cells of either α or a mating type are converted in this way to diploid $\alpha\alpha$ or aa cells. Such diploid cells are sterile and no longer sporulate or only occasionally sporulate at low frequency ($< 1\%$).

A second problem that has been encountered is that stable genetic changes from one mating type to the other have been observed. If this occurs in a haploid population, hybridization or conjugation between the converted mating type and the original mating type results in diploid cells. Such diploids are likely to be different from those obtained when the mating is conducted with a normal strain of opposite mating type. These disadvantages have been overcome, however, by the use of biochemical mutants in which the haploid populations are labeled not only by mating type but, in addition, by inducing one or more nutritional deficiences (auxotrophic markers), such as a requirement for a purine or some amino acid. For example, a mutant that requires adenine for growth can be crossed with another haploid strain of opposite mating type that requires L–arginine. This will yield a hybrid that can grow in a minimal medium devoid of both adenine and L–arginine. But if the diploid had resulted from selfing or mating-type mutation, it would not grow in the minimal medium.

MATING TYPES IN HOMOTHALLIC YEASTS

A homothallic yeast is one in which all four spores, when allowed to germinate separately, will create sporulating diploid cultures. A number of *Saccharo-*

myces species and those of many other genera belong to this category. Although it was thought at one time that such yeasts lacked mating types, it is now known that even these yeasts carry the same *a* and *α* mating type alleles as do heterothallic yeasts.

Recently, explanations for homothallic behavior have been proposed. For example, in several strains of *Saccharomyces* two independent homothallic genes, HO_α and HM, have been discovered. These genes have no linkage with the mating type locus (*a* or *α*) or with each other. The function of HO*α* is that of a gene which promotes interconversion of the *α* mating type allele to *a*, thereby creating a stable inheritable change. This has been found to occur shortly after spore germination—for example, after two cell divisions. The remaining *α* cell can then fuse with the newly formed *a* cell, giving a diploid (and sporogenous) *aα* cell. However, conversion of an *a* mating type allele to *α* during germination of an *a* spore requires the presence of both HO*α* and a modifier gene HM. Thus, the *a* and *α* mating type alleles in *Saccharomyces* are interchangeable with each other due to a highly specific mutagenic action of the homothallism genes. This change occurs at extremely high frequency, probably 100% of expectation.

Similar kinds of homothallism genes have been found in *Kluyveromyces lactis*. In this homothallic yeast the mating type alleles *a* and *α* (not necessarily corresponding to the alleles with the same designation in *Sacchyaromyces*) are subject to the influence of the respective genes, H_a and H_α. The former effects the change of *a* to α, and the latter does the converse. Undoubtedly similar systems are present in other homothallic yeasts.

CELLS OF HIGHER PLOIDY

Cells with ploidy higher than diploid have been constructed and are occasionally found in nature. We have seen how cells homozygous for mating type can arise by selfing or by endomitosis. When mixed with haploid α cells, diploid *aa* cells will conjugate and form triploid $aa\alpha$ cells. Similarly, diploid $\alpha\alpha$ cells will produce triploid $\alpha\alpha a$ cells when mixed with cells of mating type *a*. Mixing of diploid *aa* cells with $\alpha\alpha$ cells will result in conjugation and the formation of tetraploid $aa\alpha\alpha$ cells. Pentaploid lines have been obtained by conjugation of $\alpha\alpha a$ cells with *aa* cells or $aa\alpha$ with $\alpha\alpha$ cells. Finally, hexaploid cells may be obtained from the conjugation of $aa\alpha$ and $\alpha\alpha a$ cells. The cell's average DNA and RNA content, its volume, dry weight, and soluble protein have a linear relationship to genome number (ploidy). Spore viability in triploid yeasts is often poor. But their viability in tetraploid cells is high, which allows them to be studied satisfactorily. The spores derived from the latter are diploid, and depending on the segregation of the mating type alleles, they may be *aa*, $\alpha\alpha$ (incapable of sporulation), or aα (able to sporulate). If a tetraploid hybrid carries a gene B in the duplex state (B/B/b/b), the diploid spores are expected to segregate 4:0 (all spores are B/b), 3:1 (B/ B, 2B/b, and b/b), and 2:2 (2 B/B and 2 b/b). The relative frequencies of the three classes depend on how closely gene B is linked to its centromere and the orientation of the four chromosomes during the first meiotic division; such data provide information on the location of that gene.

SEGREGATION OF PHENOTYPIC PROPERTIES IN DIPLOID YEASTS

Much work has been done on the segregational behavior of genetic traits by observing individual ascospores of various yeast hybrids and of naturally occurring diploid yeasts. In many cases when a fermenter of a certain disaccharide was hybridized with a nonfermenter of that sugar, thus producing a diploid yeast, the four ascospores showed a 2:2 segregation upon sporulation of the hybrid. In such a segregation two of the ascospores in each ascus of the hybrid produced cultures that were able to ferment the sugar, and the other two were unable to do so. This indicated that a single gene was responsible for the fermentation of the sugar or, more precisely, responsible for synthesis of an enzyme necessary for the hydrolysis of the disaccharide sugar involved.

When one of the conjugating partners can ferment two sugars (represented, for example, as F_1F_2) and the other partner cannot do so (f_1f_2), the hybrid ($F_1f_1F_2f_2$) will ferment both. If the hybrid sporulates, the abilities to ferment the two sugars are often assorted independently from each other, each one in a 2:2 ratio. In some cases certain combinations of characters segregate as if they were one unit. This means that the genes responsible for these properties are located close together on the same chromosome, a phenomenon called "linkage." For example, in *Saccharomyces* the genes M_1 and R_1 (two genes responsible for the fermentation of maltose and raffinose, respectively) move as a single unit during reduction division and hybridizations.

A major controversy followed, however, when investigators observed that upon analyses of many asci, in some hybrids, certain asci segregated 2:2, others 3:1 (3 fermenters to 1 nonfermenter), and still others segregated out 4:0 (in which all four ascospores were able to ferment the particular sugar). Winge and Roberts showed that in some cases the fermentation of certain sugars, such as maltose, may be controlled by more than a single gene (polymeric genes). They concluded that their strain of *Saccharomyces cerevisiae* contained at least three nonlinked polymeric genes—M_1m_1, M_2m_2, and M_3m_3—for the production of maltase. The term "nonlinked" signifies that these genes move independently during meiosis. Allowing for the phenomenon of crossing over between homologous chromosomes and the independent assortment of these three genes, good correlation was found between the statistically expected and actual numbers of asci, giving 2:2, 3:1, and 4:0 segregations upon experimental analysis. As a result, Winge and Roberts concluded that segregation in yeasts with polymeric genes follows Mendelian laws. However, not all species contain polymeric genes for the fermentation of maltose. For example, *Saccharomyces italicus* contains only a single gene for the fermentation of maltose, and during sporulation only 2:2 segregations are obtained.

Winge and Roberts proved their hypothesis of polymeric genes through an elaborate series of hybridizations. By means of crossings and back crossings they were able to isolate strains of a specific genotype that contained only a single active maltase gene—for example, $M_1m_1m_2m_2m_3m_3$, $m_1m_1M_2m_2m_3m_3$, or $m_1m_1m_2m_2M_3m_3$. They showed that any member of a set of polymeric genes is adequate for enzyme function. Later, additional nonlinked M genes were found, so that at least six maltase genes are now known in *Saccharomyces*. Polymeric genes also occur with respect to the fermentation of several other sugars—for example, sucrose and raffinose. The significance of these genes has become clearer in recent years. Halvorson and collaborators found that the proteins representing the various maltases controlled by the different M genes apparently have similar properties. Also, a yeast with a single M gene produces less maltase enzyme than a yeast with two or more M genes. In fact, the effect of gene dosage on enzyme production was approximately additive.

Reduction division (meiosis) has also been followed cytologically by staining techniques. Staining clearly demonstrated that the four densely stained masses of chromatinic material formed from the original diploid nucleus are the meiotic products. Occasionally some of the nuclei may undergo one or more additional divisions, and ultimately eight or sometimes as many as ten or twelve nuclei are formed. This may result in asci with five to twelve spores as has been observed in certain strains of *Saccharomyces* species. These extra divisions of the nuclei are termed "supernumerary mitoses." A result of this phenomenon is that occasionally an ascospore may contain two nuclei. For example, a yeast of the genotype M_1m_1 undergoes reduction division. Initially, four nuclei are produced—two M_1 and two m_1. Suppose that the two nuclei characterized by M_1 divide once more, and that the two extra M_1 nuclei become associated with the m_1 nuclei. If these pairs become incorporated into the spores formed, the result would be two haploid spores, each containing an M_1 allele, and two "diploid" spores containing paired M_1 and m_1 nuclei. This creates an irregular segregation, since all spores upon isolation would now give clones that ferment maltose. On the premise of the original yeast, which was heterozygous for maltase, a 2:2 segregation would be expected. These exceptions are fortunately rather rare. They can be recognized with proper genetic markers that reveal the diploid condition.

Another interesting phenomenon related to the problem of irregular segregation is the property of yeast that clones with a ploidy higher than $2n$ can be formed. It is now known that some irregular segregations observed by various investigators must be ascribed to the occurrence of yeast of higher ploidy among the population.

Finally, the phenomenon of gene conversion must

be considered, first invoked by Lindegren as an explanation of irregular segregation in yeast, and later studied in detail by Mortimer. Briefly, gene conversion differes from crossing over during meiosis in that during the latter process a reciprocal recombination of genetic markers occurs, whereas in the former the process is nonreciprocal. Mortimer found that the frequency of irregular segregation due to gene conversion differed for various allelic pairs. For most loci (allelic pairs), segregations averaged approximately 1% 3:1, 98% 2:2, and 1% 1:3, although for some loci the percentages of 3:1 and 1:3 were significantly higher. In the absence of gene conversion, the expected ratio would have been 100% 2:2 segregations. The mechanism of gene conversion is still unknown, and the exact relationship between conversion and crossing over remains uncertain.

LINKAGE OF GENES AND CONSTRUCTION OF CHROMOSOME MAPS

Genetic maps are important not only to basic genetic studies, but also for one's understanding of the numerous physiological processes and their control taking place in a cell. Ascomycetous yeasts are convenient subjects for genetic analysis because of the possibility of spore tetrad analysis. Such analyses reveal information on the type of genetic control of certain traits (complementary or polymeric genes), the linkage between genes, and the linkage between genes and the centromeres of the chromosome on which they are located. Information obtained from such studies is then used to construct genetic maps.

Meiotic mapping. Suppose we have a diploid yeast that is heterozygous for two traits; A and B represent the dominant genes from one parent, and a and b are the recessive alleles from the other parent. Tetrad analysis of the four ascospores in many asci will yield information on the position of the two genes with respect to each other and their linkage to their respective centromeres. This technique, which is termed "meiotic mapping," involves a determination of the relative fequencies of three ascal classes in a number of asci. There are three possible classes of tetrads from such a hybrid as shown here:

Parental ditype (PD)	Nonparental ditype (NPD)	Tetratype (T)
AB	aB	AB
AB	aB	Ab
ab	Ab	aB
ab	Ab	ab

If the two genes are located very close to each other on the same chromosome (closely linked) so that no crossing over between the two genes can occur, only PD asci are found. When the two genes are somewhat farther removed from each other, a limited number of crossovers or exchanges can occur. A *maximum* of one crossover between A and B leads to a T ascus, and no NPD asci are found. If the two genes are still farther apart, a second crossover may occur and all three ascal classes may be found depending on the number of chromatids involved in the two exchanges. If two strands are involved a PD ascus results, with three strands a T ascus, and with four strands a NPD ascus. The relative frequencies of double-crossover ascal classes expected on random involvement of strands is PD:NPD:T of 1:1:2. Two genes are considered linked if the ratio of PD to NPD asci is significantly greater than one. At the other extreme, when the two genes are located far apart on the same chromosome so that many exchanges can occur between them there is essentially independent assortment and the PD:NPD:T ratio is 1:1:4.

The distance between genes is expressed in centimorgans, the standard genetic map unit. The length of a genetic interval in centimorgans equals 100 times the average number of crossovers in that interval per chromatid. If the proportion of NPD asci is small in comparison with PD asci, and assuming there are two or less exchanges between the linked markers, their distance X in centimorgans can be calculated from the proportion of ascal classes by the formula

$$x = \frac{100\,(\mathrm{T} + 6\ \mathrm{NPD})}{2(\mathrm{PD} + \mathrm{NPD} + \mathrm{T})}$$

By comparing different markers in relation to each other, a genetic map showing the distances between genes can be constructed. Such calculations are accurate only if the interval between genes is relatively small. The only way of accurately measuring long intervals is by the summation of shorter intervals. If there is a maximum of one exchange within the interval (that is, there are no NPD asci), the distance in centimorgans between the two genes is equal to one-half the percentage of T asci.

If A and B are located on different chromosomes, assortment of the two genes results from random disjunction of the respective centromeres at the first meiotic division, as well as from exchanges between each of the genes and their respective centromeres. If both genes are closely linked to their centromeres (thus no crossovers between genes and centromeres), the PD:NPD:T ratio equals 1:1:0. Tetratype asci will occur only if there has been a crossover between at least one of the genes linked to its centromere. In this case the PD:NPD:T ratio will be 1:1:<4. If at least one of the two genes is not centromere linked (thus if one or both genes exchange freely with their centromeres), the ratio becomes 1:1:4. Centromere linkage of a gene is usually detected by including known centromere-linked genes (such as *trp l*, which is about 1 centimorgan from its centromere) in the cross. The distance of a gene to its centromere can be measured by the frequency of second-division segregation (which is analogous to the frequency of T asci of two-gene markers). The distance in centimorgans is equal to one-half the percentage of second division segregation if the interval between the gene and its centromere is small. If two genes are near their centromeres and if at least one of these is involved in crossover between it and its centromere, the following relation holds

$$T = w + y - 3/2wy$$

in which T is the frequency of tetratype asci and w and y are the second-division segregation frequencies of the two genes.

The effect of linkage between two genes or between genes and their centromeres can be summarized in the following way:

	Ratios of ascal classes		
	PD	NPD	T
Random assortment (= independence). Two genes far apart on the same chromosome *or* on different chromosomes and at least one gene nonlinked to its centromere:	1 :	1 :	4
Centromere linkage (genes on different chromosomes and both near their centromeres):	1 :	1 :	<4
Linkage between two genes on the same chromosome:	1 :	<1	

Mitotic mapping. Linkage studies are also possible by the technique of mitotic mapping. Mitotic recombination in heterozygous diploid strains results in homozygosity due to reciprocal exchanges of genes located distal to the point of the event on the chromosome as shown here:

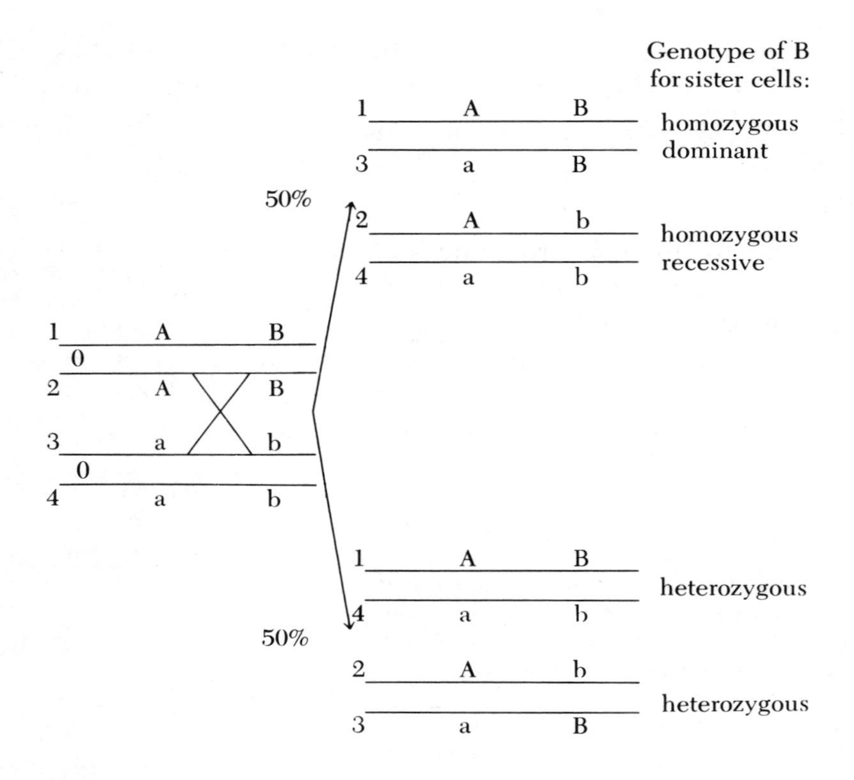

Since the spontaneous rate of mitotic recombination is very low, it is customary to use a mild treatment with mutagen to increase the rate. As shown here, after cell division there is a 50% chance that one of the cells will create a sectored colony that is homozygous for the distal B locus. If gene b represents an adenine requirement fostering red pigment production, the recessive b/b locus will result in a pink sector in the ensuing colony. The frequency of mitotic crossing over of a gene is expected to increase approximately linearly with the gene's distance from its centromere. For example, Roman has found that the frequency of spontaneous mitotic segregation of *ade–3* in *Saccharomyces* was approximately ten times that of *ade–6* on chromosome VII.

Lindegren in 1949 was the first to identify three independently assorting centromere-linked genes, *ade–1*, *gal–1*, and α, by which he provided evidence for at least three chromosomes. Subsequent research has revealed that the minimum number of haploid chromosomes (n) in *Saccharomyces*, as determined by genetic mapping, is seventeen. Hundreds of genes have been located on the genetic map of this yeast. There is evidence that the chromosome number in haploid representatives of other genera of yeast is smaller. But mapping in other genera is less well developed than in *Saccharomyces*.

RESPIRATION-DEFICIENT MUTANTS

A very special type of biochemical mutation involves the loss of respiratory activity. Many kinds of yeast, including all species of *Saccharomyces*, possess both a respiratory pathway of metabolism (substrate combustion by oxygen consumption) as well as a fermentative pathway (alcoholic fermentation). (See Chaper 6.) Respiration deficiency that prevents a yeast from growing on a nonfermentable substrate (for example, ethanol or glycerol) may result from either a nuclear gene mutation or a mutation of genes present in mitochondrial DNA. In both types of mutations the phenotype on glucose-yeast autolysate agar consists of a slow-growing, small colony (termed *petite*), resulting from less energy being generated from fermentation than from respiration. Respiratory-sufficient yeasts form larger colonies *(grandes)*.

When haploid respiratory-deficient (RD) mutants caused by nuclear mutations are crossed with respiratory-sufficient wild-type haploids, the hybrids form asci whose spores consist of two *petite* and two *grande* segregants. Such mutants, which are usually obtained by UV radiation, generally lack only a single respiratory enzyme (for example, cytochrome C) that is coded for in the nuclear DNA. Such mutants are called "genic" or "segregational," and they may revert to wild type occasionally.

The other types of RD yeasts are often referred to as "cytoplasmic" or "vegetative" mutants. The fundamental investigations of Ephrussi and his colleagues have shown that in a normal population of baker's yeast, approximately 1% of the cells are abnormal in that they do not have the ability to respire. When isolated, these spontaneously arising mutants were found to be completely stable. Further work by Ephrussi and later investigators has shown that this mutational event results in the simultaneous loss of a number of respiratory enzymes of mitochondrial origin and coded for by mitochondrial DNA. Furthermore, when haploid cultures are mutated in this way and mated with wild-type cells of the opposite mating type, the mode of inheritance is non-Mendelian. The genetic symbol for such mutants is ρ^- while respiratory-sufficient cells are designated ρ^+. The rate and extent of mutation can be greatly increased by certain chemicals, such as acriflavine and ethidium bromide. The latter compound, in a concentration of 10 μg/ml, is especially effective. In one hour under appropriate conditions and with suitable yeast strains, 100% of the population may be converted to ρ^- cells, and after twenty-four hours almost all cells become ρ° or neutral *petites*. Some investigators believe that ethidium bromide stimulates a mitochondrial nuclease that causes mitochondrial DNA breakdown. A short treatment of baker's yeast with ethidium bromide causes a deletion of part of the mitochondrial DNA, and the remaining DNA of the ρ^- culture possesses higher mol % A + T(AT). For example, the AT content of the wild-type mitochondrial DNA may go up from about 75% to 90%. Upon continued treatment of the cells with ethidium bromide, all of the mitochondrial DNA is eventually lost in an irreversible manner, giving ρ° or neutral *petites*. When haploid neutral *petites* are crossed with ρ^+ cells of opposite mating type, all zygotes produced during a mass mating are respiratory sufficient. Thus, the ρ^+ cells supply the normal mitochondrial DNA required for respiratory function in the diploid hybrid. When such a hybrid is allowed to sporulate, all four spores are ρ^+ (non-Mendelian segregation), indicating that the mitochondrial DNA after its replication in the diploid is distributed during meiosis in the four ascospores.

The situation is different, however, when ρ^- mutants are used that contain an altered form of mitochondrial DNA. Zygotes from a mass mating of such ρ^- cells with ρ^+ cells usually give a mixture of ρ^+ and ρ^-

diploid cells. Thus in contrast to neutral *petites*, ρ^- strains with altered DNA may interfere with the maintenance or development of the normal ρ^+ character in the diploids formed. ρ^- mutants of various origin differ in the degree of this interference. This property of ρ^- strains is termed the "degree of suppressiveness," which is defined as the probability that a zygote resulting from the cross of a ρ^- mutant and a normal ρ^+ strain will create a clone composed entirely of ρ^- diploids. ρ^- strains vary from highly suppressive (although never reaching 100%) to a low degree of suppressiveness. In ρ° neutral *petites* the percent suppressiveness is zero. Recent evidence has shown that when suppressive *petites* are crossed with normal ρ^+ cells, the mitochondrial DNA of the hybrid has a buoyant density intermediate between that of the two parents. This evidence, together with evidence based on the inheritance of certain drug resistance markers (which are coded for in the mitochondrial DNA), has shown that mitochondrial DNA undergoes recombination during hybridization of ρ^- and ρ^+ strains. Suppressiveness can therefore be thought of as the formation of hybrid DNA that is still lacking in essential genes coding for respiratory enzymes.

Studies of respiration deficiency and other aspects of genetic investigations in yeast have gone far beyond what is possible to cover in this short space. It is remarkable that in forty years yeast genetics has developed into a highly specialized science that has made extremely important contributions of fundamental importance to biology and biochemistry.

6 / Metabolic Activities of Yeast

Although most yeasts are not as diversified as bacteria in their metabolism, various species of yeasts may exhibit appreciable differences in metabolic potential by the various carbon and nitrogen compounds they can utilize and by the resulting metabolic end products.

Yeasts are most commonly thought of as fermentative organisms, since they can carry out the well-known alcoholic fermentation, expressed by the conversion of 1 mole of glucose into 2 moles each of ethanol and CO_2. The reason for this common notion is probably because so much work has been done with baker's yeast and brewer's yeast—both strongly fermentative species of *Saccharomyces*. However, *Saccharomyces cerevisiae* also possesses the potential for respiratory utilization of glucose, as well as other compounds that cannot be fermented (for example, ethanol and glycerol). If low concentrations of glucose are used (for example, less than 0.1%), baker's yeast has the ability in the presence of air to respire or oxidize the glucose to CO_2 and water. Depending on the conditions of growth, *S. cerevisiae* can therefore shift its metabolism from a fermentative to an oxidative pathway; both systems provide energy to the cell for growth, although the latter yields almost ten times as much as the former. It has been estimated that about three-fourths of the energy of anaerobic metabolism is lost in the form of heat. The reason for the greater efficiency of respiration, in part, is that the ethanol accumulating during anaerobic fermentation is itself a substrate for respiration. Pasteur was the first to demonstrate that a *Saccharomyces* yeast, fermenting low concentrations of glucose, decreases its fermentative activity when subjected to aeration, and part of the glucose is respired to CO_2 and water. This phenomenon, which bears the name "Pasteur effect," has received a practical application in the production of baker's yeast. In this process (see Chapter 10) the yeast is grown under intense aeration to obtain high cell yields and to minimize ethanol production.

Another phenomenon that affects the respiratory capacity of baker's yeast is the "glucose effect," or the "Crabtree effect," which in essence means that if a yeast has been growing under anaerobic conditions in a moderately concentrated glucose solution (for example, 5%), aeration of the medium does not result in respiratory activity. Instead, the yeast population continues to ferment the glucose and carries on "aerobic fermentation." With lower glucose concentrations, some respiratory activity may be observed during aeration. In general, the lower its concentration, the greater is the proportion of glucose respired. The explanation most commonly accepted is that in the presence of high concentrations of glucose, inhibition of the respiratory enzymes' synthesis occurs by the catabolic products of the readily utilizable carbon source glucose; this is "catabolite repression." Only after the glucose concentration drops to a low level does the induction of the respiratory enzymes begin. This phenomenon is also considered in baker's yeast production by keeping the glucose concentration low during growth of the yeast. The enzymes responsible for fermentation are constitutive, therefore, a yeast grown under respiratory conditions is able to ferment glucose without delay. There are some species of yeast that do not exhibit the Crabtree effect—for example, species of the genus *Kluyveromyces* (such as *K. fragilis* and *K. marxianus*) and some haploid species of *Saccharomyces* (such as *S. rosei*). These yeasts, as far as they have been studied, are also *petite*-negative (see Chapter 5)—in other words, they are unable to produce respiratory-deficient (ρ^-) mutants.

There are also many yeast species that lack the ability to ferment sugars. Such yeasts are dependent entirely on an oxidative metabolism for energy supply and growth—for example, all species of *Rhodotorula* and *Cryptococcus* and some of *Candida, Torulopsis,* and certain other genera. In addition, there are intermediary types with a strongly respiratory and a weakly fermentative metabolism, represented by several species of the genera *Debaryomyces* and *Pichia*. The reverse is also found—namely, that a yeast has a relatively weak respiratory metabolism, but is strongly fermentative. This situation is exemplified by *Saccharomyces carlsbergensis*, a yeast used for the fermentation of lager beers. Still other yeasts (such as *Torulopsis pintolopesii*, a yeast occurring in the intestinal tract of rodents) lack a respiratory ability altogether and depend exclusively on the energy derived from fermentation.

ALCOHOLIC FERMENTATION

When glucose or fructose is added to a suspension of *S. cerevisiae*, the sugar enters the cell extremely rapidly across the plasmalemma with the aid of a stereospecific transport system. Glucose and fructose are taken up by a common transport system (at somewhat different velocities), about a million times faster than they would enter by simple diffusion. The transport of these sugars is termed "facilitated diffusion"; it goes to equilibrium, is not accumulative, and no metabolic energy is required as it is for the uptake of amino acids.

As soon as either sugar enters the cell, it is phosphorylated to the respective hexose–6–phosphate with the aid of a common enzyme hexokinase and adenosine triphosphate (ATP). After the glucose–6–phosphate is converted to fructose–6–phosphate, another enzyme phosphorylates the latter to fructose–1, 6–biphosphate with the expenditure of one more mole of ATP. This reaction is followed by several metabolic steps, via the Embden–Meyerhof pathway of glycolysis, to 2 moles of pyruvate with the generation of 4 moles of ATP, amounting to a net gain of 2 moles of ATP. The 2 moles of pyruvate are decarboxylated, giving 2 moles of CO_2 and 2 moles of acetaldehyde; the latter are subsequently reduced to 2 moles of ethanol with the aid of alcohol dehydrogenase and the reduced coenzyme nicotinic acid adenine dinucleotide (NADH) formed during an earlier intermediate step. The overall result of the degradation of one molecule of hexose is two molecules each of CO_2 and ethanol and a net gain of two molecules of ATP. This form of high-energy phosphate is used to supply the energy required for cell growth and synthesis of storage products (for example, glycogen and trehalose). However, even in the absence of growth (for example, in a glucose-containing medium but without a nitrogen source), the yeast cells will convert approximately 70% of the glucose to CO_2 and ethanol while the remainder is being assimilated to carbohydrate storage products. After the glucose in the medium is consumed, the cells can utilize the stored carbohydrates very slowly by endogenous fermentation. Investigators think that utilization of endogenous reserves furnishes the cell with maintenance energy for the replacement of proteins and ribonucleic acids that are constantly undergoing breakdown.

Many years ago Kluyver established three basic fermentation rules that, to a large extent, still hold. First, if a yeast does not ferment D–glucose, it will not ferment any other sugar. Second, if yeast ferments D–glucose, it will also ferment D–fructose and D–mannose (but not necessarily D–galactose). Third, if a yeast ferments maltose, it does not ferment lactose, and vice versa. Only few exceptions to the last rule have been discovered; for example, *Brettanomyces claussenii* can ferment both maltose and lactose.

The reasoning that underlies these observations is essentially as follows: First, the fermentation of di-, tri-, or polysaccharides always goes through the hexose stage. Hydrolytic enzymes at the cell surface or located internally break down these carbohydrates to hexose sugars. Thus, if hexose sugars are not fermented, neither are di- or oligosaccharides. Second, the three sugars, D–glucose, D–fructose, and D–mannose, are transported into the cell by a common transport system (facilitated diffusion) and are phosphorylated by the same enzyme (yeast hexokinase) to the respective hexose–6–phosphates. The rates of transport and of phosphorylation vary between the three sugars. They also depend strongly on the species and even strain of yeast. Both glucose–6– and mannose–6 –phosphate are then transformed into the common denominator fructose–6–phosphate. D–galactose, however, is transported by a separate transport system that is inducible (rather than constitutive as for the other hexoses). After entry into the cell, it is phosphorylated to galactose–1–phosphate by a different enzyme galactokinase. The further metabolism of the phosphorylated galactose involves three additional enzymes:

$$\text{Gal-1-P} + \text{uridine diphosphate glucose (UDPG)} \rightarrow \text{GLU-1-P} + \text{UDPGal}$$

This reaction is catalyzed by the enzyme uridyl transferase. UDPG is regenerated from UDPGal by the action of UDPG–4–epimerase, while glucose–1–P is converted to glucose–6–P by phosphoglucomutase. The absence of genes coding for galactokinase and the other necessary enzymes explains why many yeasts that can ferment glucose are unable to ferment galactose.

Enzymes for the hydrolysis of sucrose, melibiose, raffinose, inulin, and starch are located outside the plasmalemma, but their precise location is not known. Some investigators believe them to be present in the periplasmic space—that is, between the plasmalemma and the alkali-insoluble glucan—whereas others favor the hypothesis that they occur in the mannan layer, because most or all of these enzymes also are mannan–protein complexes (glycoproteins). Other sugars, such as maltose, lactose, cellobiose, and melezitose, are hydrolyzed inside the cell after being transported across the plasmalemma by specific and inducible carriers, called "permeases." The maltose carrier is best known; the enzyme is specific for maltose, and its synthesis in a yeast previously grown on glucose (where the permease is lacking because its synthesis is repressed) requires energy. After induction of the maltose permease, the maltose entering the cell induces the synthesis of maltase that splits the maltose into two molecules of glucose. When the fully induced cell is then grown in glucose, the permease is repressed first, followed by the disappearance of maltase.

Although the number of species of yeast that can ferment hexoses, disaccharides, and trisaccharides is large, yeasts that are able to ferment polysaccharides, such as inulin (a polymer of fructose) and starch, are relatively rare. *Kluyveromyces fragilis* is an example of a yeast that ferments inulin well, and *Saccharomycopsis (Endomycopsis) fibuligera* and *Saccharomyces diastaticus* are examples of yeasts that ferment soluble starch moderately to slowly. Sugars most commonly fermented by yeasts are glucose, galactose, maltose, sucrose, lactose, trehalose, melibiose, and raffinose. Yeasts have also been found that can ferment melezitose (a trisaccharide), the disaccharide cellobiose, or α–methyl–D–glucoside. Relatively little information is known about the ability of yeasts to ferment the more rarely found di- and trisaccharides, such as those with β–(1→3)–, β–(1→6)–, α–(1→3)– or mixed linkages. No yeasts have ever been found that can ferment or grow anaerobically on pentose sugars or methyl pentoses, although many yeasts can utilize these compounds very well by a respiratory pathway. In this connection note that sugars such as cellobiose, which are not normally fermented well, often are good substrates for growth through the respiratory process.

The trisaccharide raffinose (see structure in Chapter 1) can be fermented in several different ways. Most sucrose-fermenting yeasts, such as *Saccharomyces cerevisiae*, hydrolyze sucrose with the aid of β–fructosidase (invertase), and such species also split off the fructose portion of raffinose. This results in a 1/3 fermentation of the molecule, and melibiose is left as a residue. However, not all sucrose-fermenting yeasts can ferment raffinose— for example, *Saccharomyces italicus*, which lacks β–fructosidase but hydrolyzes sucrose by the enzyme α–glucosidase. This linkage cannot be hydrolyzed by the enzyme in raffinose because α–glucosidase access is blocked by the presence of galactose. Complete (3/3) fermentation of raffinose occurs by using yeasts (for example, *Saccharomyces kluyveri)* that possess both β–fructosidase and α–galactosidase (melibiase). There are several species of yeast (for example, *Saccharomyces eupagicus)* that possess these two enzymes but lack the ability to ferment galactose (2/3 raffinose fermentation).

Occasionally, hydrolytic enzymes can act as transferases of sugar moieties. For example, *Sporobolomyces singularis* is a yeast that grows on lactose but cannot utilize the galactose moiety of this sugar as an energy source. In a lactose medium at pH 3.7 this yeast utilizes the glucose portion for growth, but the β–galactosidase (lactase) transfers the galactose moiety to unused lactose to form a trisaccharide. By an additional galactosyl transfer reaction, part of the trisaccharide is converted to a tetrasaccharide. Thus, instead of being transferred to water (hydrolysis), a hydroxyl group of a sugar serves as an acceptor of the glycosyl moiety. Other nonutilizable sugars added to a lactose medium (for example, L–arabinose) also can act as acceptors. A number of unusual di- and trisaccharides have thus been synthesized with the aid of enzymes from *Sporobolomyces singularis*. The reactions are

E (enzyme) + lactose (substrate)→
E–galactosyl + glucose

E–galactosyl + lactose (acceptor)→
galactose–galactose–glucose + E

E–galactosyl + $(\text{galactose})_2$–glucose→
$(\text{galactose})_3$–glucose + E

The maximal concentration of ethanol that a yeast can produce by fermentation depends strongly on the species and even on a particular strain of a species. Suitable strains of *Saccharomyces cerevisiae* (for example, a wine yeast or distiller's yeast), given a sufficient supply of sugar, can obtain an alcohol concentration 12–14% by volume relatively rapidly. But beyond this alcohol level the rate of fermentation quickly decreases and finally becomes very slow. Fermentations of grape and fruit juices have been recorded to reach alcohol levels of 18–19% by volume. Some strains of *Schizosaccharomyces pombe* also can produce relatively high levels of alcohol, but species of the apiculate yeasts (for example, *Kloeckera apiculata),*which are often abundant during the early stages of natural wine fermentation, are not able to surpass a level of more than 3.5–6.0% alcohol. The fermentations by yeasts that potentially can produce high levels of alcohol are generally rather unpredictable; very high levels of alcohol can be achieved only with certain yeast strains after many months of slow fermentation at a reduced temperature (approximately 14 –17°C) and with sugar concentrations adjusted to a low level.

An interesting property of some yeasts is their ability to ferment L–malic acid to ethanol and CO_2:

$$COOH.CH_2.CHOH.COOH \rightarrow CH_3CH_2OH + 2CO_2$$

Although some strains of *Saccharomyces cerevisiae* can carry out this reaction with low efficiency, certain fission yeasts, such as *Schizosaccharomyces pombe* and *Schizosaccharomyces malidevorans* are much more effective. Little is known about the biochemistry of this conversion. Under aerobic conditions, the proportion of ethanol is reduced and that of CO_2 is increased. There is considerable industrial interest in this reaction in connection with the conversion of L–malic acid in wines of low sugar and high levels of fixed acids.

RESPIRATION

The respiratory activity of yeast has been studied over many decades by some of the outstanding biochemists of the century— for example Warburg, Wieland, Keilin, Krebs, Kornberg, and many others. The respiratory pathways that are known to exist in yeast include the citric acid cycle, or tricarboxylic acid (TCA) cycle, and its more recently discovered variation, the glyoxylate cycle. The different enzymes that catalyze the various steps of the cycle are located in the mitochondria. The TCA cycle starts with pyruvate, which, instead of being decarboxylated to acetaldehyde as during alcoholic fermentation, is oxidized to acetyl—coenzyme A. Acetyl–CoA may also arise from fatty acid oxidation. This compound condenses with oxaloacetate, giving citrate, the first compound of the actual TCA cycle. During the subsequent reactions of the cycle, two molecules of CO_2 are released from the substrate. The various oxidation and reduction reactions of cycle intermediates are not only involved in the generation of ATP by oxidative phosphorylation, but several intermediates also serve as precursors for the synthesis of cellular constituents.

The glyoxylate cycle or glyoxylate bypass serves to replenish TCA cycle intermediates (in particular L–malate and succinate) lost because of biosynthetic reactions. It also allows growth of yeasts on substrates with two carbon atoms (for example, ethanol and acetate). Since the TCA cycle eliminates two molecules of CO_2, growth on such substrates would not be possible without replenishing the lost carbon atoms. A second replenishing reaction known to occur in yeast is the biotin-dependent reaction between CO_2 and pyruvate to oxaloacetate by the enzyme pyruvate carboxylase. Kornberg has introduced the term "anaplerotic pathways" for those, such as glyoxylate cycle, fulfilling this replenishing role.

A third cycle known to exist in yeast is the pentose cycle or the "hexose monophosphate shunt mechanism," in which glucose, after being phosphorylated in the usual way to glucose–6–P, is then oxidized to 6–P–gluconate with the reduction of NADP to $NADPH_2$. During a subsequent step an additional molecule of $NADPH_2$ is generated, concomitant with the release of one molecule of CO_2. Under aerobic conditions, operation of the full cycle can result in the complete oxidation of one molecule of glucose, from which are generated six molecules of CO_2 and twelve molecules of the (reduced) coenzyme $NADPH_2$ rather than ATP. However, part of the $NADPH_2$ can be oxidized via transhydrogenation and the respiratory chain, thus contributing to the generation of ATP. Much of the reduced $NADPH_2$ is used, however, in biosynthetic reactions that require reductions (for example, lipid synthesis), since, on the average, the yeast constituents are in a more reduced state than the nutrient sugar. In baker's yeast grown aerobically 6–30% of the glucose is thought to be respired via the pentose cycle. However, in other yeast species the percentage may be higher. This cycle, as well as the TCA cycle and the ancillary glyoxylate cycle (whose details can be found in the standard biochemistry textbooks) have an important bearing on the production of certain by-products of metabolism.

Although much of the initial work on respiratory activity in yeast has been done with hexose sugars, a greater array of substrates can be respired than fermented. In contrast to many bacteria, pentoses and methyl pentose sugars cannot be fermented by yeasts. However, species endowed with the proper pentose kinases and other enzymes needed to convert these sugars into intermediates of the pentose cycle can respire them efficiently. Some species of *Cryptococcus* and *Rhodotorula* often grow more vigorously on such *compounds as* L–arabinose, D–xylose, and L–rhamnose than they do on glucose. Other compounds that can only be respired include sugar alcohols (glycerol and glucitol), organic acids (lactic, succinic, and citric acids), aliphatic alcohols (methanol and ethanol), hydrocarbons, and even aromatic compounds (phenol, cresols, catechol, and benzoic acid). Few yeasts can degrade the last category of compounds. However, strains of *Trichosporon cutaneum* can utilize many aromatic compounds, whereas *Candida tropicalis* is more selective in the aromatics it can degrade. The ability of yeasts to grow on hydrocarbons (n–alkanes) is found among many species of *Candida* and *Torulopsis. Saccharomycopsis lipolytica, C. tropicalis,* and *C. maltosa* show excellent growth on n–alkanes; those with twelve to eighteen carbons appear to be utilized best. Among the ascomycetous yeasts, many species of *Debaryomyces, Pichia, Metschnikowia,* and *Schwanniomyces* assimilate n–alkanes well, but such ability is lacking among species of *Saccharomyces, Kluyveromyces, Hansenula, Lipomyces,* and *Dekkera.* Growth of yeasts on methanol has received much attention in recent years because this simple one-carbon substrate is considered useful as a substrate for single cell protein production (see Chapter 10). Relatively few yeasts can utilize this compound, but many of those that do show excellent growth— for example, *Pichia pastoris, Hansenula polymorpha, Candida boidinii,* and *Torulopsis sonorensis.*

The ability or inability of yeast to utilize representative carbon compounds for growth has been widely used as a basis for taxonomic differentiation of species (see Chapter 11). Metabolic reactions involving a chain of several enzymes that occur before an intermediate of metabolism enters the established metabolic cycles are apparently of greater value in this connection than those reactions requiring a single enzyme, and are thus controlled by a single structural gene.

STORAGE CARBOHYDRATES

The biochemistry of storage carbohydrate formation from various substrates is rather complex. These processes are covered comprehensively in various handbooks and reviews and are covered only briefly here. The two storage carbohydrates in yeast are glycogen and trehalose (see Chapter 3). When glucose is the substrate, part of it is stored as these products during metabolism. Under appropriate conditions, they can be broken down again and utilized as energy sources. The synthesis and breakdown of the two carbohydrates take place by different pathways. The following reactions summarize the synthesis of glycogen:

$$(1)\quad \text{glucose–6–P} \xrightarrow{\text{phosphoglucomutase}} \text{glucose–1–P}$$

$$(2)\quad \text{glucose–1–P} + \text{uredine triphosphate (UTP)} \xrightarrow{\text{pyrophosphorylase}} \text{UDPG} + \text{PP}$$

$$(3)\quad \text{nUDPG} + (\text{G})_{m} \xrightarrow[\text{transferase}]{\text{glucosyl}} \text{nUDP} + (\text{G})_{m+n}\ \text{(linear)}$$

$$(4)\quad (\text{G})_{m+n}\ \text{(linear)} \xrightarrow[\text{enzyme}]{\text{branching}} (\text{G})_{m+n}\ \text{(branched)}$$

In reaction (3) linear chains of α–(1→4)-linked glucose residues are formed, whereas in reaction (4) a branching enzyme removes a terminal section of the linear chain and connects it to a neighboring chain by an α–(1→6) bond. Both chains then elongate and the branching process repeats itself, resulting in a tree-like structure with a molecular weight of about 10^6 daltons.

The breakdown of glycogen involves the action of several different enzymes. The outer chains are degraded by phosphorylase:

$$(G)_{m+n} + \text{phosphate} \xleftrightarrow{\text{phosphorylase}} \text{glucose–1–P} + (G)_{m+n-1}$$

Phosphorylase is unable to shorten the outer α–(1→4)-linked maltodextrin chains to fewer than four glucose residues. These short remaining "stubs" are removed by a combination of two enzymes. An oligotransferase removes maltotriose and attaches it at the end of a neighboring chain, while amylo–1,6–glucosidase removes the remaining glucose residue (debranching) as free glucose. The debranched chain can then, in turn, be degraded by phosphorylase.

Similarly, trehalose is synthesized and degraded by different enzyme systems. The synthesis proceeds as follows:

$$(1)\quad \text{G–6–P} + \text{UDPG} \longrightarrow \text{trehalose–6–P} + \text{UDP}$$

$$(2)\quad \text{trehalose–6–P} \xrightarrow[\text{phosphatase}]{\text{specific}} \text{trehalose} + \text{phosphate}$$

Trehalose utilization simply involves hydrolysis of this nonreducing disaccharide into two molecules of glucose by the specific hydrolytic enzyme trehalase.

When the yeast cell is faced with the assimilation of compounds with less than six carbon atoms, entry into the various metabolic cycles occurs at different points. For example, compounds with four and five carbons are phosphorylated and enter the pentose cycle; three-carbon compounds enter the glycolytic pathway at the stage of triose–phosphate; two-carbon compounds enter the TCA cycle (often as acetyl–CoA).

In reversing the glycolytic pathway and the TCA cycle, several irreversible steps must be overcome. One of these is the phosphorylation of fructose–6–P to fructose–biphosphate. This step is bypassed during gluconeogenesis (anabolism) by the induction of a highly specific enzyme, fructose–biphosphatase, which hydrolyzes the phosphate from carbon–1. Another irreversible step is the conversion of phosphoenolpyruvate to pyruvic acid. For example, a yeast growing on lactic acid first converts this substrate to pyruvate. The latter is then converted by pyruvate carboxylase and CO_2 to oxaloacetate; a second reaction leads to phosphoenolpyruvate by the enzyme phosphoenolpyruvate carboxykinase. Each of these steps requires expenditure of high energy phosphate. Enzymes involved in these anabolic reactions have been termed "gluconeogenetic enzymes."

BY-PRODUCTS OF METABOLISM

Since qualitatively and quantitatively the by-products of metabolism may vary greatly under anaerobic and aerobic conditions, we discuss them separately under the two environmental conditions.

Under fermentative conditions approximately 95% of the glucose (excluding cellular growth) is converted to ethanol (48.4%) and CO_2 (46.6%). *Saccharomyces cerevisiae* also produces small amounts of glycerol, succinic acid, higher alcohols (fusel alcohols), 2,3–butanediol, and traces of acetaldehyde, acetic acid, and lactic acid. Glycerol is quantitatively the most important by-product and results from the enzymatic reduction of some of the dihydroxyacetone phosphate to glycerol phosphate, and then by phosphatase action to glycerol. Most of the glycerol is formed during the early stages of fermentation after adding the inoculum. When the conditions of fermentation are strictly anaerobic, very little succinate is produced; however, in the presence of air and in the final stages of fermentation, a small amount of succinate accumulates in the fermentation broth, probably with the participation of the glyoxylate bypass of the TCA cycle.

The accumulation of higher alcohols during fermentation has been studied extensively, probably because of the effect of these alcohols on the quality of alcoholic beverages (especially distilled ones). Higher alcohols are produced in small amounts by all strains of yeast that have been investigated. The major components are isoamyl, optically active amyl, isobutyl, and n–propyl alcohols; a minor component is n–butanol. These alcohols were originally thought to originate exclusively from amino acids by the Ehrlich mechanism. This involves the removal of the amino group by transamination; the resulting keto acid is then decarboxylated and reduced to an alcohol with one carbon atom less than in the original amino acid:

$$\underset{\text{COOH}}{\overset{\text{R}}{\text{HCNH}_2}} \xrightarrow{\text{transamination}} \underset{\text{COOH}}{\overset{\text{R}}{\text{C=O}}} \xrightarrow{\text{decarboxylation}}$$

$$\underset{\text{H}}{\overset{\text{R}}{\text{C=O}}} + CO_2 \xrightarrow[\text{(NADH}_2\text{)}]{\text{reduction}} \overset{\text{R}}{\text{CH}_2\text{OH}}$$

Examples are the conversion of leucine to isoamyl alcohol and isoleucine to optically active amyl alcohol. Later findings, however, revealed that higher alcohols were also formed when ammonium ion was the nitrogen source. This observation—together with the fact that certain fusel alcohols (for example, n–propanol and n–butanol) correspond to α–aminobutyrate and norvaline, respectively, neither of which occur naturally—led to the view that these higher alcohols are formed (at least in part) from 2–keto acids generated by carbohydrate metabolism. Some of these keto acids are precursors in amino acid biosynthesis, whereas others are unrelated to it. These keto acids can be decarboxylated and reduced as shown in the preceding equation.

Under aerobic conditions of growth, a number of by–products have been found in surprisingly high concentrations. The nature and concentration of these products varies greatly with the species of yeast and cultural conditions used.

Species of the genera *Brettanomyces* and *Dekkera* produce high concentrations of acetic acid from glucose or ethanol. They apparently cause an aerobic fermentation of glucose; the ethanol formed then undergoes an incomplete oxidation to acetic acid. The concentration of this acid may become so high that the yeast is rapidly killed off unless adequate buffering by calcium carbonate is provided. Acetic acid is also produced in large amounts by certain species of *Hansenula* (such as *Hansenula anomala*), but here the acetic acid becomes enzymatically esterified with ethanol to form ethyl acetate. This yeast can synthesize ethyl acetate in very high yields from glucose or alcohol, provided the pH of the medium is not too high. In *H. anomala*, ethyl acetate is a transitory by–product, because upon longer incubation and aeration this ester is used up by the yeast. Many esters have been detected and identified in media inoculated with different species of yeast. In most cases these esters, which are mainly responsible for the various odors of fermented culture media, are present in rather low concentration. The technique of gas liquid chromatography has helped greatly in their identification.

Under aerobic conditions, the concentration of succinic acid is greatly increased. This can be attributed to the participation of the glyoxylate bypass of the TCA cycle when alcohol is being oxidized to acetyl–CoA. The latter compound is converted to isocitrate, which in turn yields succinate and glyoxylate. Glyoxylate does not accumulate but is converted to L–malate after reacting with another molecule of acetyl–CoA.

Another nonvolatile acid is produced by *Kloeckera brevis* (and probably other species of *Kloeckera*), *Debaryomyces hansenii, Hansenula subpelliculosa, Trichosporon captitatum,* and several others. This acid , which has been named zymonic acid, is a cyclic tetronic acid. In the presence of sufficient calcium carbonate, some of these yeasts can convert about half of the glucose utilized to this acid:

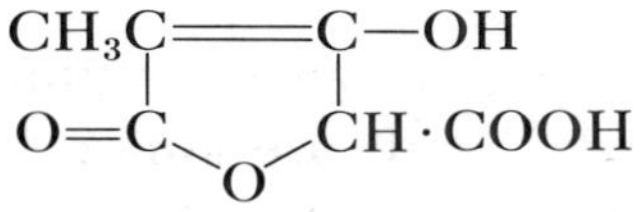

$$\begin{array}{l} CH_3C{=\!=}C{-}OH \\ \quad\;\; | \qquad\;\; | \\ O{=}C \diagdown_{O} \diagup CH\cdot COOH \end{array} \qquad \text{Zymonic acid}$$

Many of the haploid species of *Saccharomyces* and the imperfect yeasts belonging to *Torulopsis* and *Candida* can produce very high yields of polyhydric alcohols from glucose. These include glycerol, erythritol, D–arabitol, and D–mannitol. The formation of these compounds is strongly enhanced under highly aerobic conditions; anaerobically, ethanol and CO_2 are the main products and the polyol concentration is reduced. These compounds are related to the intermediate products of the pentose cycle. For example,

$$\text{D–ribulose–5P} \xrightarrow{\text{phosphatase}} \text{D–ribulose} \underset{}{\overset{\text{NADPH}}{\rightleftharpoons}} \text{D–arabitol}$$

The last (reductive) step is catalyzed by a specific dehydrogenase enzyme using NADPH as coenzyme. The formation of these products varies qualitatively and quantitatively with the species of yeast (see Chapter 10). Low levels of inorganic phosphate stimulate production of these polyhydric alcohols, probably because of increased phosphatase activity (derepression).

Very interesting extracellular lipids or glycolipids are formed by a number of yeast species in aerated cultures. The glycolipids can be observed as crystals or as a heavier-than-water oily liquid, both of which collect at the bottom of the flask after the aerated medium comes to rest. Four different types of lipids have been structurally identified:

1. Polyol fatty acid esters, in which saturated, unsaturated and hydroxy fatty acids are joined by ester linkages to C_5 and C_6 polyols. The latter were usually mixtures of D–mannitol, D–arabitol, and xylitol. Besides saturated and unsaturated fatty acids, the mixture of glycolipids also contained 3–D–hydroxyhexadecanoic acid and 3–D–hydroxyoctadecanoic acid (the carboxyl carbon of the fatty acid is termed C_1). One molecule of a long-chain fatty acid is attached to each polyol molecule, and many of the remaining hydroxyl groups, including the one on the hydroxy fatty acid, are acetylated. These compounds are formed by certain strains of *Rhodotorula glutinis* and *Rhodotorula graminis*.

2. Hydroxy fatty acid glycosides of sophorose, a sugar in which two glucose molecules are linked by a β–(1→2) bond. In such compounds a hydroxyl group of a fatty acid is glycosidically linked to the reducing carbon (C_1) of the sugar. A 13–hydroxydocosanoic acid (C_{22}) glycoside is produced as needle-shaped crystals by *Candida bogoriensis*. The sophorose moiety also

contains two acetyl groups. A second sophorose glycoside is produced by *Torulopsis apicola*. In this compound (a heavy oil) the fatty acid moiety was either 17–L–hydroxyoctadecanoic or 17–L–hydroxyoctadecenoic acid. It also contained two acetyl groups per molecule. *T. apicola* also excretes a neutral lipid in which the carboxyl group of the fatty acid is esterified with a hydroxyl group of the sophorose.

3. An unidentified species of yeast has been shown to produce still another kind of extracellular lipid,—erythro–8,9,13–triacetoxydocosanoic acid, a trihydroxy fatty acid with twenty-two carbon atoms.

4. Another class of lipid materials is produced by certain species of *Hansenula*—those which form mat, dull colonies and pellicles in liquid media. This class is composed of acetylated sphingosines or sphingolipids, which were studied extensively in *Hansenula ciferrii*. In suitable strains of this yeast, extracellular microcrystals of sphingolipids are found among the cells. Between 1–2% of the glucose supplied can be transformed into sphingolipid under optimal conditions. The most abundant component is tetraacetylphytosphingosine, which is formed as follows:

$$CH_3(CH_2)_{13}\cdot CHOH\cdot CHOH\cdot CHNH_2\cdot CH_2OH$$

$$+$$

$$\text{4 acetyl–CoA}$$

$$\downarrow$$

$$CH_3(CH_2)_{13}\underset{|}{\overset{OCOCH_3}{CH}}\text{———}\overset{OCOCH_3}{\overset{|}{CH}}\text{———}\overset{NH\cdot COCH_3}{\overset{|}{CH}}-CH_2OCOCH_3$$

Tetraacetylphytosphingosine

A minor component of yeast sphingolipid has been identified as triacetyl C_{18}–dihydrosphingosine. These hydrophobic compounds are thought to be involved in the formation of surface pellicles in liquid media.

Most species of *Cryptococcus* excrete an amylose-like polysaccharide, but only when the pH of the medium of growth decreases to a level of about 3.0 or below. The latter is usually accomplished by growing the yeast in a mineral medium with ammonium sulfate as the nitrogen source. The formation of amylose is readily demonstrated by the blue color formed when iodine solution is added to the medium, or by a blue halo around colonies growing on agar plates. A few species, such as *Cryptococcus macerans*, produce this starch-like compound independent of a lowering of the pH. The biochemical mechanism of this synthetic process is not well understood, but it could be an aberration in glycogen synthesis—an impairment of the branching process.

PIGMENTS OF YEAST

Colonies of most species of yeast appear white, or white with a slight greyish or yellowish cast. Carotenoid pigments of various pink and yellow shades are produced by species of the following genera: *Sporobolomyces, Sporidiobolus, Rhodosporidium, Rhodotorula, Phaffia,* and some species of *Cryptococcus*. All of these are asexual or sexual genera related to the class *Teliomycetes*. Thus far, no ascomycetous yeasts have been found that contain carotenoid pigments.

Rhodotorula species produce sufficient pigments, so that the colonies assume various shades of red, depending on the pigment composition (often very bright in color). Carotenoids of *Rhodotorula* (or *Rhodosporidium*) species are composed mainly of two oxygenated compounds, torulene and torularhodin. The concentrations of these two major pigments relative to each other vary greatly in different species and strains, and result in different shades of red.

A salmon-red carotenoid, astaxanthin, is the major pigment of *Phaffia rhodozyma*. This pigment, which is similar to that found in lobsters and trout, is not present in other red yeasts.

Most species and strains of *Cryptococcus* synthesize small amounts or no carotenoids at all. As a result, their colonies are often only slightly yellowish or pinkish, or even colorless. However, some strains of *Cryptococcus laurentii*, especially when grown on certain carbon compounds in mineral media, produce rather high concentrations of β–carotene, resulting in bright yellow colonies.

Minor concentrations of other carotenoids occur in most of the pigmented yeasts mentioned here—for example, α–carotene and lycopene. The last pigment is well known from its prominence in tomato fruits. The synthesis of the carotenoids starts with acetyl–CoA, which is converted to a basic five-carbon branched terpenoid (isopentenyl pyrophosphate) in a series of reactions. This intermediate is then converted to a C_{40} compound, and in the final reaction the C_{40} chain is rearranged and altered into specific carotenoids. The structure of β–carotene is shown as an illustration of a carotenoid pigment:

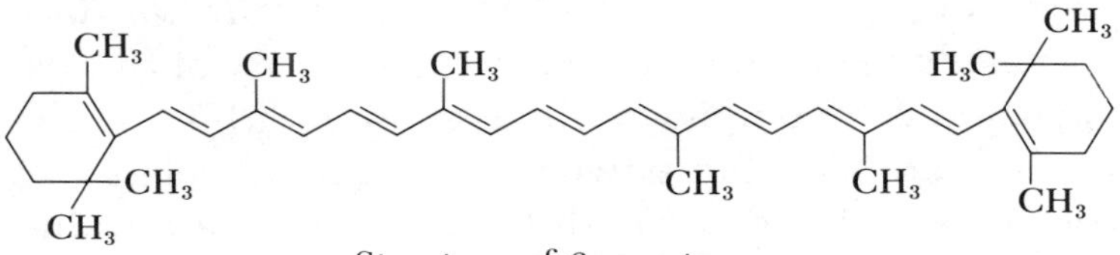

Structure of β–carotene

The absolute concentration of the carotenoid pigments in yeast is always low. The yellow colored strains of *Cryptococcus laurentii* var. *flavescens* contain about 6–10 mg of β–carotene per 100 g dry weight of yeast cells. The pink species of *Rhodotorula* have been reported to contain 3–17 mg per 100 g, of which 15–43% was β– plus γ–carotene, the remainder being torulene and torularhodin.

A striking effect of environmental conditions on pigment production has been observed, but these effects may vary greatly with the yeast strain or species. For instance, *Rhodotorula rubra* had essentially the same reddish color when grown at 25°C as at 5°C, but a particular strain of *Rhodotorula glutinis* was salmon-pink when grown at 25°C and a dark yellow after growth at 5°C. Another species, *Rhodotorula peneaus* (now considered synonymous with *Cryptotoccus laurentii* var. *flavescens*), was canary-yellow at 25°C but cream colored at the low temperature.

The amount of pigment formed is strongly depressed in some strains when grown in aerated cultures; in other strains, aeration has little influence. Some oxygen is obviously required, since not only are species of these genera strict aerobes, but colonies growing on a plate show a much higher pigment concentration at the surface than in the deeper layers. Finally, the carbon compound of a synthetic medium upon which these yeasts are grown affects the color and pigment composition of certain strains, whereas others are only slightly or not at all affected. Mutations affecting pigment synthesis have also been noted.

In the past, variation in pigmentation of carotenoid-containing yeasts was used as a basis to establish varieties and in some cases even species. As a result of the considerations discussed here, pigmentation no longer carries any weight in the differentiation of species and varieties, although in generic diagnoses it still has limited value.

Not all pinkish colonies of yeast owe their color to carotenoid pigments. It has long been known that upon streaking of the yeast *Metschnikowia (Candida) pulcherrima*, the colonies may assume a deep maroon-red color, which occasionally diffuses into the medium, forming a reddish halo around the growth. At the beginning of the century Beijerinck had already established that this pigment was insoluble in organic solvents (in contrast to the carotenoid pigments), was slightly soluble in water, but could be extracted with alkali and reprecipitated with acid. We now know that the formation of this pigment, which has been named pulcherrimin, is dependent upon the presence of iron in the medium. Pulcherrimin is a symmetrical macromolecule of high molecular weight, containing 12.7% iron in addition to carbon, hydrogen, oxygen, and nitrogen. Its unit structure may be regarded as the iron salt of pulcherriminic acid, a pyrazine derivative. Pulcherrimin is also produced by heavily sporulating species of *Kluyveromyces, Hansenula,* and *Pichia*, and by some strains of *Saccharomyces* growing in biotin-deficient media. In these cases, the spores appear to be pigmented.

Certain adenine-requiring mutants of *Saccharomyces cerevisiae* or of *Schizosaccharomyces pombe* form deep red or pink colonies. The pigment accumulated by such mutant cells is polyribosylaminoimidazole, to which are attached a number of different amino acids.

Little detailed knowledge is available on the dark pigments of the "black yeasts" that are usually placed in the genus *Aureobasidium (Pullularia)*. Usually, colonies are initially white. After a few days they go through a pinkish or greenish stage, finally becoming deep brown or jet black. The pigment apparently is highly polymerized and related to the melanins; it is insoluble in water and organic solvents.

Other groups of metabolic by-products are capsular polysaccharides (phosphomannans and various heteropolysaccharides), enzymes, sterols, and vitamins. Under certain conditions, these metabolites are overproduced to such an extent that they are excreted into the medium of growth. (For further details, see Chapters 3 and 10.)

7 / Nutrition and Growth

The vast majority of yeasts can be cultivated under relatively uniform conditions on common laboratory media, but a few species exhibit unusual growth requirements. The latter group of organisms have highly specialized habitats that often reflect their need for special growth factors, and this requires adjustment in the overall medium composition and incubation conditions.

The usual growth requirements of yeast include an organic carbon compound, which serves as a source of carbon and energy, organic or inorganic nitrogen for the synthesis of proteins and nucleic acids, various minerals (including compounds furnishing trace elements), and frequently a mixture of vitamins.

Complex Media

Because of the close relation between the developmental history of yeast and the field of brewing, one of the oldest classical media used for the cultivation of yeast is malt extract or beer wort. It is made from diastatic malt (See Chapter 10) by enzymatic conversion of the starch and proteins to fermentable sugars and amino acids plus peptides, respectively. The solution thus formed contains all of the required ingredients and can be used in liquid form or as a solid medium after the addition of agar. In some countries it is available as spray-dried powder or syrup; these can be dissolved or diluted with water to a concentration of 1–10% soluble solids, depending on what is desired. If malt extract is not available, it can readily be made in the laboratory by preparing a slurry of 20% diastatic malt in water and heating it for about fifteen to twenty minutes at approximately 65°C. After enzymatic conversion it is filtered, and a malt extract solution with about 15% soluble solids is obtained.

Malt extract is sometimes modified by the addition of yeast extract and peptone. Wickerham recommends yeast extract/malt extract agar (YM) for maintaining or storing yeasts (see Table 1).

In general, yeasts are not affected appreciably by changes in pH, and nearly all species can grow within a wide range of pH values. We have found that few, if any, yeasts are inhibited by a pH value of 3.0, provided the acidification is done with hydrochloric or phosphoric acid. *Nematospora coryli* no longer grows when the pH in such a medium is lowered to 2.8, although most yeast species can tolerate a pH that low. However, at low pH, certain organic acids, such as lactic and acetic acids, are inhibitory to the growth of most yeasts. This is apparently due to the penetration of the undissociated form of the acids across the plasmalemma into the cytoplasm. At the other end of the pH scale, yeasts readily grow at values of 7–8 and often even higher, but optimum growth is normally found somewhere in the range between pH 4.5 and 6.5.

Yeast autolysate is another commonly used complex medium; it contains all of the necessary growth requirements with the exception of a *fermentable* carbon source. ("Fermentable" is stressed since many yeasts can grow aerobically to a greater or smaller extent at the expense of the available amino acids in yeast autolysate, but fermentation and gas production are not possible with amino acids.) Yeast autolysate (often referred to as yeast extract) is made by incubating a baker's yeast slurry at 55°C (usually with the addition of a little toluene) for about three days. During this period the cells are killed, but the various enzymes of the cell (such as proteases, nucleases, phosphorylase, and trehalase) are activated. They cause breakdown of the proteins, nucleic acids, glycogen, and trehalose—a process termed "autolysis." During the early stages of autolysis, the cell's trehalose and glycogen are fermented to ethanol and CO_2; hence, the final filtered product no longer contains any fermentable carbohydrate. Yeast autolysate is therefore an ideal medium with which to study sugar fermentation reactions of yeast after supplementing the basal autolysate solution with the desired sugar. Yeast autolysate (or extract) is normally available as a spray-dried powder.

SYNTHETIC MEDIA

In many growth experiments synthetic media are used. Two widely used synthetic media, developed by Wickerham, are given in Table 2. These very useful media are sold commercially in prepared form. Yeast Nitrogen Base (YNB, Difco) contains a number of trace elements, nine vitamins, trace amounts of amino acids to stimulate growth of certain fastidious yeasts, and the principal minerals—potassium phosphate, magnesium sulfate, sodium chloride, and calcium chloride. The nitrogen source is ammonium sulfate, which can be utilized by most known yeasts. The desired carbon source must be added, normally in a concentration of 0.5–1%.

The carbon sources that can be utilized by yeast vary greatly with the species. Glucose is utilized by all yeasts, although it is not necessarily the most effective carbon source for all species. Categories of carbon sources that various yeasts can assimilate include hexose sugars, di-, tri-, and polysaccharides, pentose sugars, methyl pentoses, the lower aliphatic alcohols, sugar alcohols (polyols), organic acids, and some miscellaneous compounds, such as aromatics (see Chapter 6). Use is made of the ability or inability of yeasts to assimilate various carbon compounds in the differentiation of species.

In Yeast Carbon Base (YCB, Difco) medium, glucose provides carbon, but a source of nitrogen is lacking. This medium can be used to determine which nitrogen sources a yeast can or cannot use for growth. Categories include single amino acids, purine and

TABLE 2. Composition of two chemically defined media for growing yeasts (amounts are given per liter of distilled water)

Ingredients	Yeast nitrogen base	Yeast carbon base
Carbon source	*grams*	*grams*
D–glucose	none[a]	10
Nitrogen source		
$(NH_4)_2SO_4$	5.0	none[b]
Salts		
KH_2PO_4	1.0	1.0
$MgSO_4 \cdot 7H_2O$	0.5	0.5
NaCl	0.1	0.1
$CaCl_2 \cdot 2H_2O$	0.1	0.1
Amino Acids	*milligrams*	*milligrams*
L–histidine · HCl · H_2O	10	1.0
DL–methionine	20	2.0
DL–tryptophan	20	2.0[c]
Compounds supplying trace elements	*micrograms*	*micrograms*
H_3BO_3	500	500
$CuSO_4 \cdot 5H_2O$	40	40
KI	100	100
$FeCl_3 \cdot 6H_2O$	200	200
$MnSO_4 \cdot 1H_2O$	400	400
$Na_2MoO_4 \cdot 2H_2O$	200	200
$ZnSO_4 \cdot 7H_2O$	400	400
Vitamins		
Biotin	2	2
Calcium pantothenate	400	400
Folic acid	2	2
Inositol	2000	2000
Niacin	400	400
Para-aminobenzoic acid	200	200
Pyridoxine · HCl	400	400
Riboflavin	200	200
Thiamine · HCl	400	400

[a] The desired carbon source must be added.

[b] The desired nitrogen source must be added.

[c] The nitrogen contained in these three amino acids is insufficient to support visible growth.

pyrimidine bases, amines, urea, nitrate, nitrite, and ammonia. Although reports have appeared in the literature that some yeasts are able to assimilate (fix) nitrogen from the air, critical experiments with isotopically labeled nitrogen have given no support for this notion. As was true for carbon sources, yeast species vary greatly in their ability to satisfy their nitrogen demand from these sources—with the exception of urea and ammonium ion, which are suitable for practically all yeasts. The suitability of individual amino acids as nitrogen sources depends on the yeast's ability to deaminate a particular amino acid and to incorporate the nitrogen into other nitrogenous constituents of the cell. Glutamic and aspartic acids are examples of amino acids that are easily deaminated (or transaminated) by most yeasts and are therefore good nitrogen sources. Normally, if a yeast can utilize nitrate, it is also able to utilize low concentrations of nitrite as a nitrogen source. Nitrite, however, is quite toxic, especially if used in too high a concentration. All species of the genera *Hansenula* and *Citeromyces* can utilize nitrate and nitrite. Some species of *Debaryomyces* are known to use nitrite, but not nitrate; this may be attributed to the lack of an enzyme to reduce nitrate to nitrite.

The utilization of urea as a nitrogen source is quite interesting. Yeasts have been classified by various authors as "urease-positive" and "urease-negative" species. The first category, which includes mainly yeasts related to the basidiomycetes (for example, *Rhodosporidium* and *Cryptococcus*), produce large amounts of free ammonia when cultured on urea as the nitrogen source. This is usually demonstrated by including an indicator in the medium (for example, phenol red, which turns a deep magenta color when ammonia causes the medium to become alkaline). Such species are thought to contain large amounts of the classical urease, which is believed to split urea directly (via carbamate) into two molecules of ammonia and one of CO_2. The urease-negative group comprises most of the ascomycetous yeasts. Although they grow well on urea as a nitrogen source, no free ammonia is produced. This mechanism of urea utilization has been studied extensively in recent years. The principal mechanism by which urea enters the cell is by an inducible, energy-dependent, active transport system. The cell then uses the urea by the following two reactions:

$$(1)\quad \underset{\text{urea}}{NH_2\cdot CO\cdot NH_2} + ATP + HCO_3' \xrightleftharpoons{Mg^{2+}}$$

$$\underset{\text{allophanate}}{NH_2\cdot CO\cdot NH\cdot COOH} + ADP + P_i + OH'$$

$$(2)\quad \text{allophanate} \rightarrow 2CO_2 + 2NH_3$$

Reaction (1) is catalyzed by the biotin-containing enzyme urea carboxylase, and reaction (2) by allophanate hydrolase. These two enzymes occur in close association and are components of a multienzyme complex. Their levels are under strict regulatory control by the cell.

VITAMIN REQUIREMENTS

All vitamins needed by yeasts, except *meso*-inositol, serve vital catalytic functions as part of coenzymes in the metabolism of yeast, regardless of whether they are supplied in the medium or synthesized by a particular yeast. Biotin, the most commonly required vitamin, is involved in carboxylation reactions (CO_2 assimilation) as shown here for the assimilation of urea (and in Chapter 6 for the conversion of pyruvate to phosphoenolpyruvate via oxalacetate). Niacin and riboflavin are components of coenzymes involved in oxidation-reduction reactions [for example, nicotinic acid adenine dinucleotide (NAD) and flavine adenine dinucleotide (FAD)]. Pantothenic acid is a component of coenzyme A and is involved, for example, in acetylation reactions. Thiamine functions as thiamine pyrophosphate in the decarboxylation of pyruvate and in rearrangement reactions of the pentose cycle. Pyridoxine is involved in transamination reactions. Folic acid, as tetrahydrofolate, is involved in the metabolism of one-carbon fragments, such as transmethylation. *p*–Aminobenzoate is a component of the folic acid molecule. Finally, *meso*-inositol serves a structural function as a component of phospholipids in membrane synthesis and stability. No yeasts are known that either require or synthesize vitamin B_{12}.

Although nine vitamins are included in the synthetic media discussed here, yeasts vary widely in their requirements for vitamins. Biotin appears to be the most commonly required vitamin, and riboflavin and folic acid the least. In fact, riboflavin probably can be synthesized by all yeasts. Many free-living species (such as *Hansenula anomala)* grow vigorously in vitamin-free media, and they synthesize all of the necessary vitamins themselves. Most species of *Saccharomyces* require one or more vitamins, although vitamin-independent strains of *Saccharomyces cerevisiae* are known to exist. Some strains of *S. cerevisiae* can convert pyridoxine into thiamine and vice versa. When such strains—if they are unable to synthesize either vitamin—are grown in a medium lacking either thiamine or pyridoxine, growth takes place. However, a double omission of both vitamins does not allow growth to take place. In tests for vitamin requirements, such a double omission is therefore routinely included.

Stating that a yeast requires a particular vitamin is often qualified by adding that the requirement is an absolute one or a relative one. Yeast having an absolute requirement for a vitamin cannot grow in the absence of this nutrilite, irrespective of the time of incubation. A relative requirement indicates that the yeast can very slowly synthesize this particular growth factor, but will grow much more rapidly if it is supplied in the medium. Yeasts with absolute vitamin requirements are useful for vitamin assays of complex substrates. For example, *Schizosaccharomyces pombe* has an absolute requirement for inositol, and it apparently is not affected by compounds with inhibitory or inositol-sparing action. All of the apiculate yeasts belonging to *Kloeckera* and *Hanseniaspora* have an absolute requirement for pantothenic acid and for inositol. Like most lactose-fermenting yeasts, *Kluyveromyces fragilis* has an absolute requirement for niacin. Some investigators have attempted a biochemical classification of yeast strains based on individual vitamin requirements, and in some genera, such as *Rhodotorula*, some of the species appear to have characteristic vitamin needs. However, because of the considerable variation in vitamin requirements between representative strains of the same species (for example, *Saccharomyces cerevisiae*), the general usefulness of specific vitamin requirements in taxonomy appears doubtful. Currently, their primary use in taxonomy is to determine whether or not a yeast is able to propagate in a vitamin-free synthetic medium.

Temperature Ranges for Growth

The optimal temperature of growth depends on the particular species with which we deal. No common denominator (a temperature at which all yeasts will grow) exists. However, for nearly all purposes a temperature of 20–25°C is best for the growth of the vast majority. Many free-living yeasts are unable to grow at 30°C. Incubators at 30°C or above are useful only for special instances.

Two categories of yeast form exceptions to this 20–25°C temperature range. There are several species, typically associated with warm-blooded animals, that have minimum growth temperatures of 24–30°C. At the other extreme, yeasts have been isolated from antarctic regions; often these species barely grow at 17–20°C, and their optimum temperature is somewhere between 12–15°C. In these examples a clear relationship between the habitats of the yeasts and their optimum temperature exists. Table 3 gives certain examples of growth ranges of yeast that we have determined in our laboratory. Obviously, there are enormous variations in the range under which yeasts are able to grow. At the lower part of a range it may take one month or longer to reach maximum cell crops due to the very low rate of growth. Maximal cell crops in the upper part of a given range are reached relatively rapidly; if at such temperatures no growth occurs after several days, one can conclude that the yeast is unable to grow at that particular temperature. The maximum temperature for growth or temperature tolerance can sometimes be raised slightly by incor-

TABLE 3. Examples of temperature ranges for growth of several species of yeast. Variations of several degrees in minimum and maximum temperatures of growth are normal for different strains of the same species.

Species	Temperature (°C)					
	0	10	20	30	40	50
Candida macedoniensis						
Debaryomyces hansenii						
Nadsonia elongata						
Leucosporidium scottii						
Cyniclomyces guttulatus						
Candida slooffii						
Saccharomyces mellis						
Schizosaccharomyces octosporus						
Pichia membranaefaciens						
Kluyveromyces fragilis						
	0	10	20	30	40	50

porating special growth factors into the medium that a yeast itself can no longer synthesize at a particular temperature. For example, *Candida slooffii* does not grow at 43°C unless the medium is supplemented with choline. This compound protects the yeast from thermal death. Death of *Saccharomyces cerevisiae* at 40°C is preventable by supplementing the medium with ergosterol plus oleic acid.

Yeasts, although they are also fungi, have lower maximum temperatures for growth than higher fungi. Several higher filamentous fungi have been found to grow as high as 55–60°C.

SPECIAL GROWTH REQUIREMENTS

Several yeast species have unusual growth requirements (referred to as exacting species). In other cases special requirements may be imposed by environmental conditions.

Although adenine-requiring mutants of *Saccharomyces cerevisiae* are well known from genetic studies, *Schizosaccharomyces octosporus* strains grow poorly in synthetic media unless supplemented with about 15 mg/l of adenine. This yeast may have arisen from a naturally occurring, adenine-deficient mutant.

We have recently discovered a new species of yeast in necrotic tissues of cacti in the North American Sonoran Desert. This yeast, which was named *Pichia amethionina*, is unable (in contrast to virtually all other yeasts) to synthesize its sulfur-containing amino acids from sulfate. Synthetic media must be supplemented therefore with approximately 10 mg/l of either

L–methionine or L–cysteine for growth to occur. All strains of this yeast, obtained over a wide geographic area, had this deficiency.

Pityrosporum ovale, which has its habitat on the scalp of humans and on the skin of certain animals, requires fats for growth. Fat is an absolute requirement; it was believed for many years that unsaturated fatty acids—in particular, oleic acid—were the required growth factors. Recently, however, by the use of highly purified fatty acids, investigators showed that *P. ovale* does not grow in a basal medium supplemented with pure oleic acid. Instead, the growth requirements can be satisfied by myristic or palmitic acid. However, oleate increases the crop of organisms in media containing limited concentrations of myristate or palmitate. This yeast apparently cannot synthesize long-chain fatty acids from C_2 units and therefore requires preformed fatty acids in the medium. Because of the immiscibility of fatty acids or lipids in liquid media at pH 4.5, growth is improved by the addition of taurocholate; even better growth is obtained in Littman's oxgall medium supplemented with fatty acids. However, these bile acids do not substitute for the higher fatty acids.

Other unusual growth requirements are found among species of the genera *Brettanomyces* and *Dekkera*. These very slow-growing yeasts are now known to require unusually high concentrations of thiamine in the medium—up to 10 mg/l, which is nearly twenty-five times as much as is required by most yeasts.

Although not a growth requirement, use is often

made of calcium carbonate as a buffer in the medium to protect those yeasts which produce large amounts of acetic acid and other organic acids. Species of *Brettanomyces* and *Dekkera* are especially short lived, because the acetic acid produced on solid media during aerobic growth tends to kill off the cells very rapidly. Incorporation of 0.5–1% calcium carbonate in the medium protects these cultures for an appreciable period against the toxic effects of acetic acid.

Several species of yeast occur as harmless parasites in the intestinal tract of warm-blooded animals. One member of this group, *Cyniclomyces* (syn. *Saccharomycopsis) guttulatus*, is a large, budding yeast, occurring in the intestinal tract of domestic rabbits (and possibly in a few other rodents). It was first observed in that habitat in 1845; but, in spite of repeated attempts, it was not until approximately 1955 that the organism was successfully cultured in the laboratory. In its natural habitat its complex nutritional requirements are satisfied by the environment and by the food being digested in the stomach of the rabbit. Growth could not be sustained in malt extract, a fairly complex medium in itself. However, growth of this yeast is possible in media containing glucose and the filtrate of thoroughly autolyzed yeast. Growth also occurs in synthetic media, provided certain protein hydrolysates are added. Suitable in this respect are Proteose Peptone, and Trypticase (a pancreatic digest of casein). It has been possible to replace the complex protein hydrolysates with a mixture of twenty-one amino acids, but exactly which amino acids are required is not known at present. Other unusual aspects of *Cyniclomyces guttlulatus* are a requirement for high concentrations of CO_2 in the atmosphere and a temperature range between 30–40°C. A gaseous atmosphere containing about 10–15% CO_2 (optimal) is especially critical when this yeast is grown on agar plates. In a liquid medium sufficient CO_2 may be generated by glucose fermentation to give adequate growth. However, if the yeast is grown near the lower end of its temperature range, sparging of the broth with a gas mixture containing 15% CO_2 and 2% O_2 in nitrogen is necessary for growth to occur at a reasonable rate. Although under otherwise optimal conditions good growth will occur at pH 2.0 (a pH close to that of the stomach contents of a rabbit), the cells become rapidly granulated and die when reaching the stationary phase. At pH 5.6 the cells are more stable and less subject to granulation, but at pH values above 7.0 no growth takes place.

Certain other psychrophobic intestinal parasitic yeasts—that is, yeasts with a relatively high minimum temperature for growth—include *Torulopsis bovina* (and its ascogenous form, *Saccharomyces telluris*), *Torulopsis pintolopesii,* and *Candida slooffii*. All of these species occur in the intestinal tract of warm-blooded animals (see Chaper 8), but they are less specific with respect to their host and less fastidious in their growth requirements than is *Cyniclomyces guttulatus*.

Some of the less exacting species, however, exhibit unusual growth requirements. For example, some strains of *T. bovina* and *T. pintolopesii* have an absolute requirement for about 1 mg/l of choline, $CH_2OH \cdot CH_2N^+(CH_3)_3$, a component of lecithin (or phosphatidylcholine). Although some of the choline in the medium appears to be incorporated into lecithin, part of it is metabolized to other compounds, possibly functioning as a methyl donor. *Candida slooffii* is the least exacting of these species. If this yeast is grown at 37°C, choline inhibits growth; but at 43°C, choline is essential in the medium, protecting *C slooffii* against thermal death. Moreover, growth of this yeast at 43°C, is vanadium dependent (about 10 μg of $NaVO_3$/l). Psychrophobic species also exhibit nutritional imbalances with respect to the concentration of various amino acids and purine and pyrimidine bases; too high a concentration of some of these media components will inhibit growth. This is probably a reflection of their highly specialized habitat in the intestinal tract of their hosts.

Another example of an environmentally imposed nutritional requirement relates to the growth of baker's yeast under strictly anaerobic conditions. Although *Saccharomyces cerevisiae* possesses both an aerobic and an anaerobic (fermentative) metabolism, growth in an otherwise complete medium (for example, malt extract or grape juice) stops after a number of generations in the complete absence of oxygen. Addition of ergosterol plus oleic acid to a culture grown under strictly anaerobic conditions allows continuous growth. Apparently, oxygen (even low concentrations are adequate) is necessary for the biosynthesis of these essential cell components. Bréchot and coworkers in France have shown that oleanoic acid (a compound somewhat related to ergosterol and not to be confused with oleic acid), which occurs naturally in the whitish bloom on grapes, also supports anaerobic growth of yeast, either in the presence or absence of oleic acid.

Some yeasts are strongly osmophilic, indicating that they thrive best in high concentrations of sugar. Most of these yeasts, as exemplified by *Saccharomyces bisporus* var. *mellis* and *S. rouxii*, are found growing in honey and other media with high sugar content. It is usually possible to adapt these osmophilic yeasts to media with low sugar content by transferring them progressively onto media with lower and lower sugar concentrations. Only rarely are osmophilic yeasts found that cannot be adapted to growth at low sugar concentrations. An example of an obligate osmophile is *Eremascus albus*, a yeast-like fungus that is a spoilage organism of foods with high osmotic pressure. This yeast requires a minimum of 40% sugar in media for it to grow properly.

The situation is somewhat different with salt. Several yeasts, especially species of the genus *Debaryomyces*, can grow slowly in nearly saturated solutions of sodium chloride, but they can be readily grown on ordinary media containing no sodium chloride at all. Such yeasts might be designated as "osmoduric," indicating that they can withstand high levels of salt but do not require them. Similarly, most yeasts isolated from sea water grow as well on media without salt as in media containing 3–4% NaCl. An exception to this rule were some strains of *Metschnikowia bicuspidata* var. *australis*, which we isolated from diseased brine shrimp grown in ponds with about 10–12% salt. These strains did not grow in YNB-glucose medium, but grew normally when 2% salt was added. A 10% malt extract medium also supported their growth. However, in general yeasts behave differently from several halophilic bacterial species that require salt for growth.

GROWTH INHIBITORS

Several antibiotics have been discovered that are very effective in inhibiting the growth of fungi (including yeasts), although they are generally inactive against bacteria. Many of these antibiotics are quite toxic when administered parenterally (however, medical applications of these compounds are not covered here).

Cycloheximide (Actidione) is an antibiotic produced by *Streptomyces griseus*. It interferes with cytoplasmic protein synthesis at the polyribosomal level. The sensitivity of yeasts to cycloheximide varies greatly. *Saccharomyces cerevisiae* and related species are inhibited completely by a few mg/l of medium, but other yeasts are highly resistant, tolerating more than 100 mg/l. Examples of the last category are *Saccharomyces montanus*, *S. florentinus*, most species of *Kluyveromyces* (for example, *K. fragilis*) and the apiculate yeasts belonging to *Hanseniapora* and *Kloeckera*.

Chloramphenicol, produced by *Streptomyces venezuelae*, on the other hand, inhibits mitochondrial protein synthesis but has no effect on cytoplasmic protein synthesis. *Saccharomyces cerevisiae,* given sufficient glucose, will grow in the presence of chloramphenicol. But after exhaustion of the glucose the alcohol formed will not be respired due to the yeast's inability to synthesize several respiratory enzymes in the mitochondria.

Lomofungin is an antibiotic produced by *Streptomyces lomendensis*. It is not only active against yeasts and fungi, but also against bacteria. *Saccharomyces cerevisiae* is inhibited by as little as 5 μg/ml. The antibiotic interferes with RNA synthesis by preventing formation of the RNA-polymerase–DNA initiation complex, probably by chelation with the firmly bound Zn^{2+} of the enzyme. If the medium contains significant levels of Cu^{2+} or Zn^{2+}, the antibiotic is inactivated.

Tunicamycin, produced by *Streptomyces lysosuperificus,* is active against yeasts and fungi as well as gram-positive bacteria. It selectively blocks the synthesis and secretion of the wall mannan–protein complex (see Chapter 3) and mannan-containing enzymes, such as invertase and acid phosphatase. Growth of *Saccharomyces cerevisiae* is not extensively inhibited, but the cells enlarge and often become spherical. Such treated cells are very sensitive to cell wall lytic enzymes.

The polyenic antifungal antibiotics form a large group of related macrolides whose ring contains a rigid planar lipophilic portion and a more flexible hydrophilic portion. They inhibit the growth of yeasts and fungi, but not of bacteria. Examples are nystatin, candicidin, amphotericin A and B, pimaricin, and filipin. The polyene antibiotics can be divided into two groups with somewhat different properties, (a) those with thirty-three to thirty-seven carbons and (b) those with forty-six or more. The polyenes are thought to exert their antifungal action by first forming a tight complex with the plasmalemma through its ergosterol component. This in turn causes altered membrane permeability, which results in a leakage of many cytoplasmic components, including proteins. Eventually, cell death occurs.

8 / Ecology

The field of yeast ecology is concerned with the manner in which yeasts live and propagate in nature, where specific organisms can be found. Conversely, it investigates the nature of a yeast population in or on substrates that can support their growth, and the reasons for the often highly specific characteristics of the various yeasts occurring in diverse natural habitats. It is also concerned with the interaction between yeasts and other groups of microorganisms—fungi, bacteria, and protozoa—as well as the higher forms of life, ranging from insects to warm-blooded animals, and from algae to the vascular plants. Most yeasts live a saprophytic life, which means that they grow on nonliving organic substrates, but parasitic types are also known. These depend largely on a living host to supply the necessary nutrients. Interaction between yeasts and other groups of microorganisms, living side by side on a particular substrate, is not necessarily limited to competition for a share of the available nutrients, although it is certainly an important factor. Several examples of forms of interaction follow:

(1). One organism may make available components of a substrate that would otherwise be unavailable to a second organism. For example, few yeasts are known that can grow on cellulose, xylan, or araban. Many molds and bacteria, however, can attack these polymers, usually by hydrolysis with extracellular enzymes. Yeasts can profit from this action by utilizing a portion of the breakdown products. We have observed that nearly all yeasts occurring in association with trees and bark beetles can utilize cellobiose, D–xylose, and often D–arabinose as single carbon sources. A similar phenomenon is starch hydrolysis by molds, followed by the development of yeasts unable to grow on starch per se but capable of utilizing the hydrolysis products. Another example involves bacterial necrosis of living plant parts by phytopathogenic bacteria. Such rotting tissue can be invaded easily by specific yeasts, usually carried by insect vectors.

(2). Some actinomycetes produce antibiotics that inhibit, in extremely low concentrations, the growth of certain fungi and yeasts. Although conditions in industry have been adjusted to obtain very high yields of these compounds, they undoubtedly are also produced in natural substrates and in concentrations sufficient to affect the composition of the yeast population. Considerable variation in sensitivity of yeasts to these antibiotics has been demonstrated. Examples are cycloheximide (Actidione) and the polyene class of antibiotics, both produced by species of *Streptomyces*.

(3). Major metabolic products of bacteria can also inhibit yeast growth or kill cells already present. Acetic acid, produced by vinegar bacteria from alcohol formed by yeasts, is an example. Many yeasts are very sensitive to acetic acid in the undissociated form, but exceptions are known (see Chapter 9).

(4). Some bacterial metabolic products of uncertain nature have been reported to inhibit yeast growth; their action does not appear to be very specific. Examples are species of several common genera of soil bacteria, a pseudomonad of marine origin, and a particular strain of *Acetobacter mesoxidans*. The antagonistic action of the last organism does not appear to be caused by acetic acid, although most yeasts are inhibited by this acid.

(5). Ingestion of yeasts by insects, protozoa and other animals occurs commonly. Such organisms consume the yeasts as food; however, they also aid in the distribution of yeasts by adherence to the exterior of their bodies, or in some cases by carrying them in special pouches or body invaginations. We have occasionally observed the rather tenacious association of protozoa in cultures of apiculate yeasts.

Since yeasts cannot be recognized with the naked eye (even 10^6 baker's yeast cells suspended in a liter of water do not create a visible turbidity), information on the distribution of yeasts in nature must depend on culturing of substrates suspected of harboring yeasts.

Even microscopic observation of natural substrates will reveal the presence of yeasts only if they are present in very large numbers. Refer again to the example where 10^6 cells were suspended in a liter of water. If a droplet of this suspension is placed on a microscope slide and covered with a coverslip, the observer would find most of the fields devoid of cells, although some might contain just a single cell. A suitable concentration for observing yeasts in this way would be about 10^9–10^{10} cells per liter.

METHODS OF ISOLATION AND STUDY

We briefly describe the methods used in the enumeration of the different species of yeasts that are present in a natural substrate, because the methods strongly influence the results. Several pitfalls in methodology soon become apparent. In spite of shortcomings, the preferred procedure is to plate the material directly on an agar medium that is likely to support the growth of the yeasts present. A limitation of this technique is that the surface of the medium in an average petri dish cannot support more than approximately 500 separate colonies. If more cells are inoculated, the resultant colonies tend to grow together. This creates difficulties, since we have to depend primarily on the appearance of colonies to make an estimate of the various types or species that develop. This, then, implies that if the population contains minority types—for example, less than 1 in 500, or 0.2%—these types are unlikely to appear on the plate. Since petri dishes inoculated with natural substrates often contain many less than 500 colonies (sometimes only a dozen), minority types could easily be overlooked. Yet it is the only method that can give a reasonable estimate of the yeasts as they are found in natural substrates.

Many studies in the past have made use of enrichment cultures in liquid media. In this procedure, a small amount of the substrate is inoculated into a liquid culture and allowed to grow for several days until the presence of yeasts becomes apparent visually, or upon microscopic inspection. This material is then plated and colonies are isolated. The weakness of this procedure is that in the artificial medium in which the substrate was inoculated, minority types could easily outgrow majority types if the former found life in the artificial medium more advantageous. Thus, a completely erroneous picture would be obtained of the actual yeasts present in the material under study. The following is a simple example: suppose there are 1,000 organisms belonging to *Cryptococcus* and a single yeast cell belonging to *Saccharomyces* in a sample of flower nectar. If this material is inoculated into a medium containing glucose and yeast autolysate and left undisturbed until growth is apparent, the population is likely to have a ratio of *Saccharomyces:Cryptococcus* of many thousands to one. Thus, upon plating and isolation, we would think that *Saccharomyces* was the predominant type, or even the only species in the original substrate. If this technique is used at all, it is best to place the sample in yeast autolysate with only 0.1–0.2% glucose in a small flask on a shaker. In this way competitive advantage of fermentative yeasts is avoided.

On solid media one of the problems that must be overcome is competition by bacteria and molds. As explained in Chapter 7, yeasts can withstand low pH values unusually well; adjustment of ordinary nutrient media to a pH between 3.5 and 3.8 (HCl is commonly used) is ordinarily sufficient to inhibit or reduce the growth of the vast majority of competing bacteria. Another approach is to incorporate in the medium one or more broad-spectrum bacterial antibiotics that do not affect yeasts.

The control of molds is more difficult. Since yeasts are also fungi and thus closely related, they frequently respond in the same way to inhibitory conditions. The most troublesome fungi are those which have a strongly spreading type of growth, as do species of *Mucor* or *Rhizopus*. Fortunately, only a few natural substrates (for example, soil samples and spoiled fruits) contain these fungi in abundance. In each ecological survey, preliminary experiments should be conducted to determine the extent to which fungal growth may be expected to interfere with the isolation of yeasts. Some antifungal agents have been used with a fair degree of success. Propionic acid is most effective at low pH values, but under certain conditions it also inhibits the growth of yeasts. The best approach is to select a borderline concentration that reduces the rate of growth of the molds and yet allows the yeasts to grow. Other investigators have used dilute dye solutions (for example, 0.003% rose bengal) or 1% ox gall incorporated in potato-glucose agar. The last two agents seem to work best in media that are not too rich in nutrients and therefore do not permit such a rapid growth of the fungi. Last, some workers have used shaken liquid cultures in which the fungal spores germinate and form small mycelial balls, preventing the formation of conidia or spores. After a limited amount of growth, the fungal mycelium can be filtered off over sterilized glass wool and the yeasts in the filtrate can be determined by conventional means. This method, however, has the inherent disadvantage of all enrichment techniques.

The media for isolating yeasts are ordinarily 5% malt extract agar, Wickerham's maintenance agar (see Chapter 7), or an agar medium containing 0.5% powdered yeast autolysate plus 5% glucose. When bacteria are a problem, the pH is adjusted to 3.7 with hydrochloric acid. Since agar breaks down when autoclaved at such a low pH, a previously determined amount of 1 N HCl is added aseptically to the medium after autoclaving and before pouring it into the petri dishes. (Special conditions as to additional growth factors, temperature of incubation, substrate concentration, and other details have been discussed in Chapter 7.) In general, it is wise to simulate the environmental conditions under which the yeast is found in its natural habitat.

Before inoculating the plates, one should consider possibly pretreating the sample and the amount of inoculum. Substrates that are in a desiccated condition (for example, a tree gum) should be rehydrated in a small amount of sterile water or in an atmosphere saturated with moisture, allowing sufficient time for proper rehydration. Insect or plant material may require dissection to avoid plating parts that may contain accidental impurities or contaminants. As an example, a number of studies have been made on the food intake of insects that feed on certain substrates that contain yeasts. To avoid isolating yeasts that accidentally adhere to the legs, body, and wings of these insects, procedures have been developed for removing the intestinal tract (the crop) and specifically plating these organs on agar plates. Rinsing the insects for one minute in 70% ethanol is also satisfactory for this purpose. It becomes even more difficult if the study involves yeasts that have their habitat inside certain specialized cells of the insect body—for example, mycetomal tissue or the epithelial layer of certain portions in the intestinal canal. The isolation of yeasts from such tissues requires considerable skill and experience.

Another problem, encountered in the study of yeasts that have been ingested by insects, is the rapidity with which these microorganisms are digested. We have shown that *Drosophila pseudoobscura,* which had been permitted to feed on baker's yeast, contained in their crops an average of 150,000 yeast cells per individual after feeding. After twenty-four hours at room temperature this number had dropped to sixty-five, and after forty-eight hours of starvation the yeast count was zero. The yeast content in an insect, therefore, depends strongly on the time of last feeding in relation to the time of dissection. For this reason, insects that are captured for a study of their microflora should be dissected as rapidly as possible; or if dissection must be delayed, they should be refrigerated to temperatures close to 0°C, where the rate of digestion is greatly reduced.

The amount of inoculum can sometimes be estimated after microscopic inspection of the material to be plated. However, if the yeast population is low in numbers, samples of various size should be tried. In time, experience will show the proper amount of inoculum to obtain a number of colonies not exceeding 300–500 per petri dish. If the yeast growth is too dense and the original substrate sample is no longer available, a sample of the densest part of the yeast growth can be suspended in sterile water and restreaked on a fresh plate.

Whenever possible, preliminary trials should be made to determine inoculum size before beginning a study of the yeasts associated with a new type of substrate. For example, yeast populations in tree exudates (slime fluxes) of most trees are relatively low in yeast concentration, and one to two loopfuls of inoculum may need to be streaked on a petri dish; on the other hand, one loopful of a cactus rot will frequently produce too many colonies. Aqueous sources, such as fresh water or sea water, require concentration of the yeasts present by passing relatively large volumes (100–500 ml) through sterile membrane filters and then placing each filter on the surface of the isolation medium for growth.

The only way to correlate the identity of colonies appearing on plates with the original population in the sample is to estimate as accurately as possible the number of different types of yeast present. Since much work is involved in the identification of individual yeast isolates, it is obviously impossible to purify and identify all colonies. Fortunately, many species exibit pronounced or subtle differences in gross morphology of the colonies, and with experience one can recognize groups or categories. Features to which particular attention is paid are variations in color, topography, degree of glossiness, texture (varying from very slimy, through pasty, to hard and tough), form or cross section of a colony, its degree of spreading or colony diameter, and the periphery (border), which may show characteristic markings or hyphal development in the form of a feathery or ciliate edge. A dissection microscope and good illumination must be used in the examination of the isolation plate to accomplish properly this task. Often the colony morphology is characteristic of certain genera or species, although the composition of the media and even the substitution of agar by gelatin may exert a profound influence. Some of these features are illustrated in Fig. 27. For example, on malt agar, red mucous colonies usually represent species of the genus *Rhodotorula;* dull, powdery, red colonies are often of the genus *Sporobolomyces;* glossy, mucous, colorless colonies could belong to the genera *Cryptococcus*, *Lipomyces*, or *Hansenula;* colonies that are cone-shaped with gently sloping sides and have a silky appearance often are members of the genus *Han-*

seniaspora; and colonies that have an extensive development of hyphal growth at the periphery and gradually change color from white to brown or black are members of the genus *Aureobasidium.*

This method of selecting colonies has its limitations, since some closely related but different species may possess almost identical colony morphologies. Conversely, yeasts with different colony morphologies may represent the same species, since mutations are not infrequent, especially in haploid yeasts. These often show up in a somewhat different colony morphology or in the form of sectors.

Valuable additional information can be gathered by taking small samples of various colonies and making microscopic observations of cell morphology, vegetative reproduction, and striking cytological features. Occasionally, one even can observe ascospore formation in the primary colonies of the isolation plate. Finally, one or two representative types of each kind of colony are brought in pure culture by the usual streaking techniques. Although isolations are usually made after two to four days of growth, plates should not be discarded at this time, since very slow-growing colonies (for example, species of *Brettanomyces)* may not appear until after one or two weeks. Some yeasts are naturally very slow growers, while others grow slowly because their cells may be in a weakened condition in the natural substrate. Some cells may be present as spores, which may require time to germinate.

In Chapter 4 we pointed out that heterothallic hap-

a

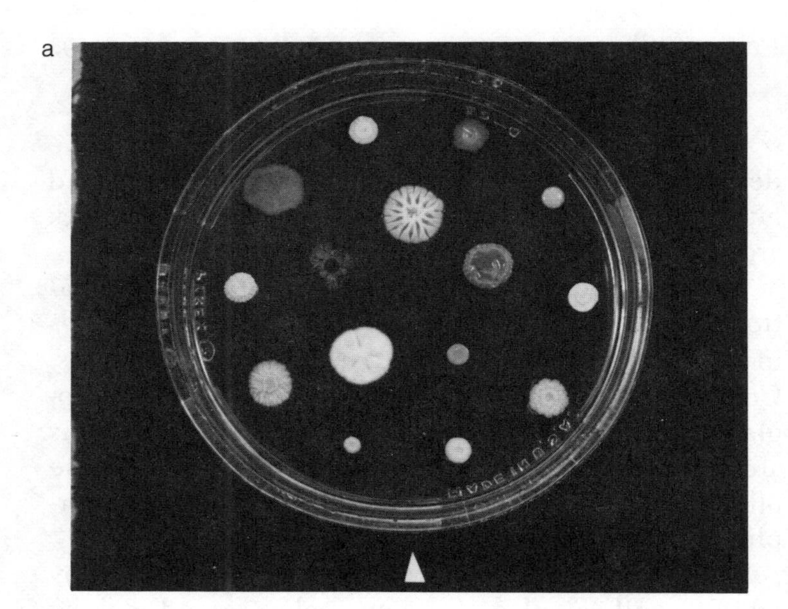

b

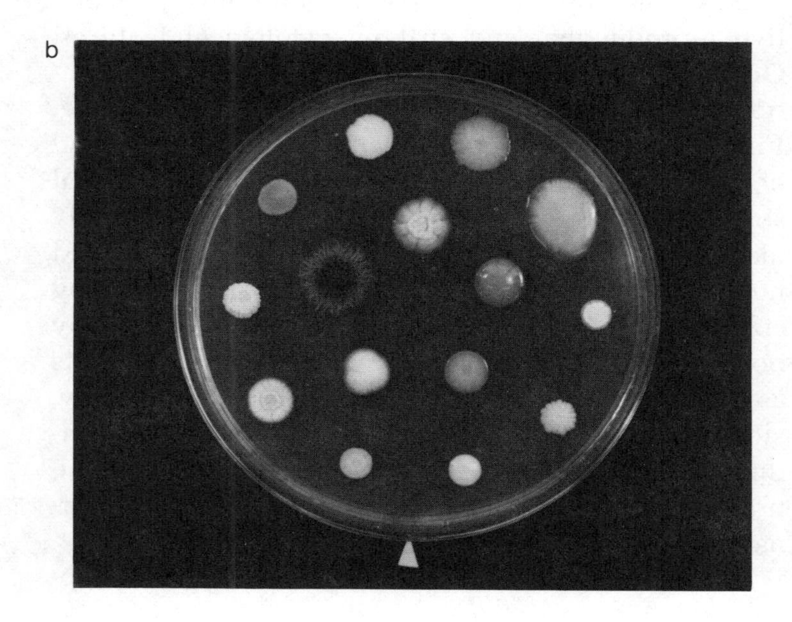

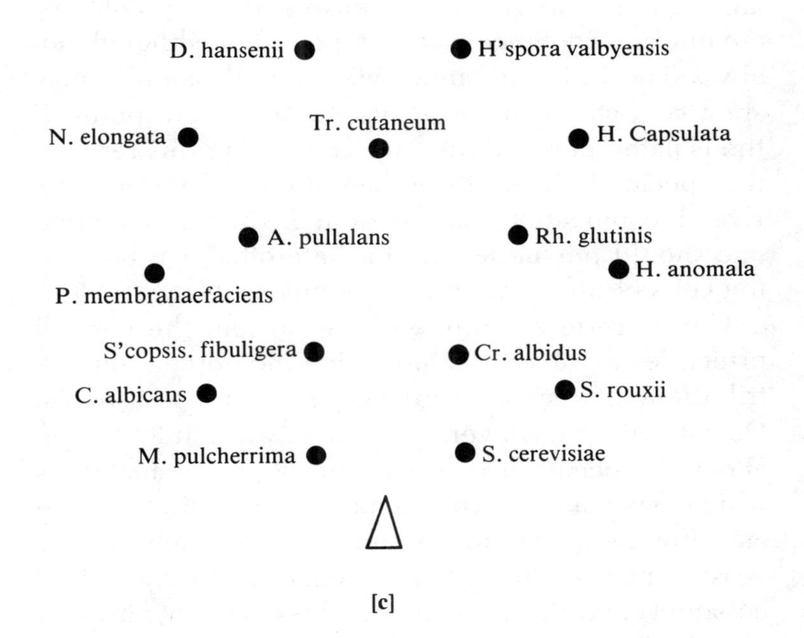

FIG. 27. Colonies of fifteen selected yeast cultures grown for eight days at 18°C in 9 cm petri dishes containing 5% malt gelatin (*a*), and for comparison 5% malt agar (*b*). The chart (*c*) is a key to the position of each culture.

loid yeasts are common in nature, which means that two mating types of such yeasts normally occur separately in natural substrates. If such a yeast is isolated by the usual purification steps, the culture generally represents the offspring of a single haploid cell. This would not be expected to produce ascospores unless it is mixed with cells of the opposite mating type. Clues of the occurrence of natural mating types can be obtained on the primary isolation plates by sampling areas where the growth is dense and many colonies are intermixed. Since mating types often (although not always) occur in the same material, such samples may show conjugation tubes or zygotes, and even spores. If the isolated pure cultures are negative in this respect, the species is likely to be haploid and heterothallic. Even if conjugation is not seen in the impure material, one should pursue testing for heterothallism by mixing cells of different nonsporogenous isolates.

The procedures outlined here present the general principles involved in elucidating the population distribution of yeasts occurring in natural substrates. Occasionally, however, one may want information about the occurrence of specific yeasts in materials which may contain a considerable excess of other species. For the quantitative enumeration of such specific yeasts, one sometimes can use selective media and incubation conditions suitable for the growth of only one or very few species, thus preventing growth of the others. The spectrum of carbon and nitrogen sources that yeasts can utilize has been discussed in Chapter 7, and application of these principles can be made in the isolation of certain yeasts from nature. A few examples illustrate this approach.

Yeasts that can utilize melibiose as a single carbon source are far less common than yeasts which are unable to use this disaccharide. By incorporating melibiose into Yeast Nitrogen Base (see Chapter 7), the selective condition is created that will allow melibiose-utilizing yeasts to develop, whereas others are unable to grow on this medium. In this way it is possible to detect and count *Saccharomyces carlsbergensis* cells (a melibiose-positive beer yeast) in a population of *S. cerevisiae*, a melibiose-negative brewer's yeast. Similar variations can be made with other carbon sources. In the same way, media can be prepared in which glucose is the carbon source (Yeast Carbon Base, see Chapter 7), and the nitrogen source is varied. If nitrate is supplied, material streaked on this medium will produce colonies only if the yeast in question can utilize nitrate. This technique has been useful in the detection of small numbers of *Candida utilis* contaminants in baker's yeast plants. *Candida utilis* is nitrate-positive and baker's yeast is nitrate-negative. Other nitrogen sources that have been suggested for differentiation include nitrite, ethylamine, L–lysine, and creatine.

Sometimes it is advantageous to use a combination of two or more selective conditions to favor a particular yeast. A case in point is the isolation of *Brettanomyces* (perfect stage = *Dekkera*) species from spoiled wines, in which such yeasts occasionally occur, but from which they are very difficult to isolate.

Van der Walt and coworkers have found that *Brettanomyces* species can be isolated with a much greater degree of success if the medium is fortified with thiamine to concentration of 10 mg/l. *Brettanomyces* species required unusually high concentrations of this vitamin. In addition, the medium is supplied with 100 mg/l of the fungal antibiotic cycloheximide (Actidione), to which *Brettanomyces* is very resistant. Wine yeasts are completely inhibited by as little as a few parts per million of this compound. These two selective conditions have made it possible to isolate cultures of the very slowly growing *Brettanomyces* yeasts from populations containing a vast majority of wine yeasts.

Another example relates to the isolation of strains of *Kluyveromyces fragilis* from natural sources. This is a thermoduric, lactose-fermenting species, and thus it is helpful to use media containing lactose as the carbon source and to incubate the plates at 45°C, since *K. fragilis* is one of the few yeasts that grows well at this relatively high temperature. Another hypothetical combination of conditions could be the following: vitamin-free synthetic medium, nitrate as the nitrogen source, i–erythritol as the carbon source, incubation at 0°C. A combination of these selective conditions certainly would eliminate the vast majority of possible competitors. Variations can also be made in the sugar or salt content of the medium, the oxygen and CO_2 concentration of the atmosphere of incubation, and other conditions. Establishment of such selective conditions, of course, presupposes an adequate knowledge of the metabolic peculiarities of a particular yeast to be isolated. However, in some instances selective conditions may not be sufficiently restrictive, so that more than one species can develop in the medium.

In selecting a suitable substrate when one wishes to isolate a certain species from nature, one should remember that yeasts are not as ubiquitous as are many bacteria. Yeasts generally have a more specific habitat. The chance of isolating a certain species, therefore, increases considerably by choosing the proper natural substrate in isolation attempts. As indicated earlier, knowledge of the distribution of yeasts in nature has been gathered from numerous surveys made by various workers. Results from these surveys indicate that some substrates are highly selective and contain only a few species of yeast (or sometimes only a single species), whereas others are much less exclusive and contain a wide variety. Even in the latter, predominant types are usually found.

In the next section we describe certain habitats and hosts that have been studied in considerable detail. These surveys, done by many different investigators, have yielded several general conclusions about the predominant yeast florae occurring in them. We also discuss where yeasts are generally found in nature. Surprisingly, yeasts occur almost everywhere, but in vastly different population densities and with great specialization for habitat. Here we stress their natural habitats.

YEASTS ASSOCIATED WITH PLANTS

Leaves

The external surface of the leaf as an environment for microorganisms has been termed the "phyllosphere." Its possible use as a habitat for yeasts and other microorganisms has been recognized only in recent years. Ruinen has shown that healthy leaves of tropical plants normally carry a rich flora of nitrogen-fixing bacteria. As leaves are also the site of photosynthesis of the plant, it is not surprising that yeasts occur in the phyllosphere, sometimes in large numbers. Leaves of elm trees and other species often exude a sugary fluid (not related to the presence of aphids) that can supply the necessary carbon source for yeast growth. However, even minute amounts of sugar, insufficient to be evident to the eye, could support many yeast cells. The work of Ruinen in the tropics and that of di Menna on pasture grasses in New Zealand has shown that only a limited number of species was present, and that the great majority belonged to the asporogenous, nonfermentative genera *Cryptococcus* and *Rhodotorula.* Also, *Aureobasidium* species (the black yeasts) were found regularly. In addition to species belonging to the last three genera, we have found fermentative yeasts, such as *Saccharomyces rosei, S. pretoriensis,* and *Kluyveromyces veronae,* in the sugary exudate of elm leaves in California. Species of *Sporobolomyces* appear commonly on the leaves and grains of cereal grasses.

The yeast flora of leaves shows seasonal variation in

numbers. It ranges between 3×10^4 and 3×10^6 per gram of pasture plant leaves during most of the year, with increases to 10^8 per gram in late summer, when 90% of the population consists of red-pigmented species. Possibly, these red, carotenoid-containing species are better protected against sunlight than are nonpigmented species. Another factor of possible significance is the presence of a slimy capsule in nearly all species obtained from leaves.

Flowers

Much work has been done on the yeast flora of flowers. It is reasonable that flowers harbor yeasts, since the nectar found at the base of the corolla of many species offers sugar upon which yeast can grow. Inoculation and transfer from flower to flower is made possible by bees responsible for pollination, wasps, butterflies, and many other insects that feed on flower nectar. Yet most investigators have found that an appreciable percentage of flower samples (usually well over 50%) contain either no yeasts or at most very few cells. On the other hand, counts of over one million yeasts per large flower, or per cluster of small flowers, have also been reported. Low counts or absence of yeasts could be caused by the very short life span of some flowers, absence of insect pollination, absence of nectar glands, and sampling too soon after a flower opens. Seasonal variations in yeast population also play a significant role, the numbers being highest in midsummer.

The yeasts reported from flowers form a fairly well-defined group. Strange as it may seem, most of the yeasts are strictly oxidative, the fermentative types forming a small minority. The black yeasts, belonging to the genus *Aureobasidium*, are perhaps most common, often comprising well over half of the total number of cells present in a sample. Species of *Cryptococcus*, *Rhodotorula*, and *Sporobolomyces* are next in abundance, followed by species of *Candida* and *Torulopsis*. As a rule very few sporulating yeasts are isolated.

Among the species of *Candida* two are particularly worth mentioning, *C. reukaufii* and the closely related *C. pulcherrima*. The first species was once known as *Anthomyces reukaufii* and later as *Nectaromyces reukaufii* (both names referring to their origin in flowers). In our laboratory we have isolated this yeast repeatedly from the flowers of the shrub *Teucrium fruticans* in California. While it grows in the nectar of flowers, this yeast, has a very characteristic thallus in the form of large cells arranged in an airplane-like shape (Fig. 28). *C. pulcherrima* can be isolated often from flowers as well as fruits—on the latter they may originate from the blossoms. Pitt and Miller have shown that *C. pulcherrima* and *C. reukaufii* have a perfect stage characteristic of the genus *Metschnikowia*, and that they are heterothallic species.

Tree Exudates

Many tree species around the world show the phenomenon of fluxing—the flowing of tree sap from a wound. The cause of this condition has been ascribed

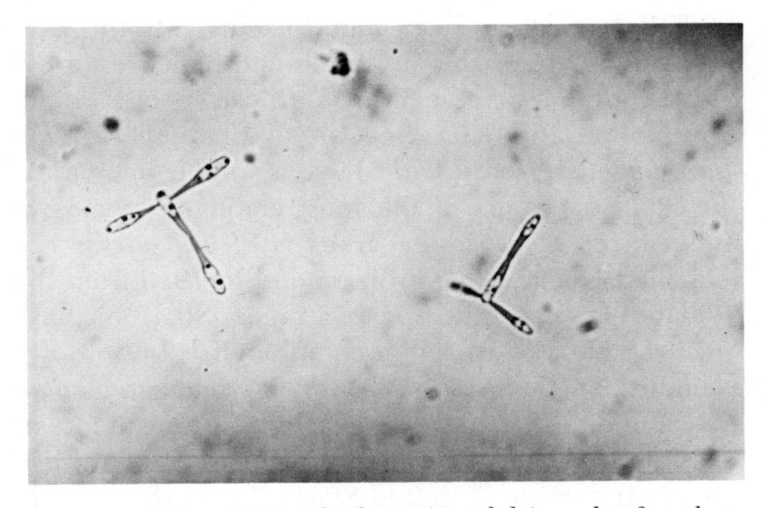

FIG. 28. Cells of *Metschnikowia(Candida) reukaufii*, photographed directly in a droplet of nectar squeezed from flowers of the shrub *Teucrium fructicans*. Note the characteristic airplane shape resulting from the aggregation of four cells.

to injuries due to boring insects or frost cracks in conjunction with bacterial infection of the wound. Tree fluxes normally persist year after year and do not appear to damage a tree appreciably, if at all. The flowing sap becomes heavily infected with bacteria and yeasts, but molds are relatively scarce. In addition, many insects find tree exudates attractive for oviposition, and larvae develop in them. The tree sap usually assumes a thick, slimy consistency (probably due to microbial polysaccharide formation), and for this reason it is often referred to as a "slime flux."

Extensive studies have been made of the yeast florae in slime fluxes. Some yeasts have their exclusive natural habitat in them. Species of *Nadsonia* and *Phaffia rhodozyma* have been found exclusively in fluxes of deciduous trees (mainly birch and beech) in areas with a cool to temperate climate (these species do not grow above 26–27°C). Species of *Endomyces, Saccharomycopsis,* and *Pichia* have a similar habitat. In our experience one of the most common species in slime fluxes of broad-leaf trees is *Pichia pastoris,* a yeast first isolated by Guilliermond in 1919 from the exudate of a chestnut tree in France. Some of these species have a restricted geographic distribution. For example, *Nadsonia* species and *Phaffia rhodozyma* are common in Japanese and eastern European forests but they have not been found in fluxes of broad-leaf trees in the western United States. In contrast, *Pichia pastoris* is common in western North America and Europe, but has not been found in Japanese forests. We think that these yeasts may be associated with specific insects which introduce them into tree fluxes and that these insects have a limited distribution. There are a number of other species of *Pichia* and of the related genus *Hansenula* that are closely dependent on deciduous trees—for example, *Pichia fluxuum*, *P. angophorae*, *P. trehalophila*, *Hansenula polymorpha*, and others. Species of *Prototheca*, a genus representing unicellular algae devoid of chlorophyll, are also very common in slime fluxes and should not be confused with yeast (see Chapter 11).

Several examples of species of other genera are: *Saccharomyces kluyveri*, which is the only species of its genus that has been isolated almost exclusively from slime fluxes of deciduous trees; *Schizosaccharomyces japonicus* var. *versatilis*, which occurs regularly in exudates of deciduous oaks in Japan and occasionally in "pine honey"; *Saccharomycodes ludwigii*, which appears to occur mainly in deciduous trees in Europe; *Citeromyces matritensis*, which has been consistently isolated from fluxes of *Myoporum* on the island of Hawaii; and *Rhodosporidium diobovatum*, which was repeatedly isolated by us from slime fluxes of deciduous trees in Japan.

Yeasts probably are introduced into slime fluxes by visiting insects. Once a yeast population is established, it usually carries only a few species and its population is quite constant. We have followed the yeast population of a particular flux of an elm tree throughout an entire year. The yeast flora was approximately constant, both qualitatively and quantitatively, consisting mainly of *Pichia pastoris*, *Trichosporon penicillatum*, and *Prototheca moriformis*. Apparently these three species had established a stable equilibrium between themselves and their natural environment. Other species, sporadically introduced by certain vectors, did not maintain themselves in this particular habitat.

Decaying stems of various cacti in the Sonoran Desert of northern Mexico, southern Arizona, and Baja California also have a distinct yeast flora. Approximately one half of the flora is represented by yeasts with phenotypic properties resembling those of *Pichia membranaefaciens*. Three of these have been named *Pichia cactophila*, *P. heedii*, and *P. amethionina*. Other abundant species are *Candida ingens* and two other species unique to cacti, *Torulopsis sonorensis* and *Cryptococcus cereanus*. Since the cactus necrosis is believed to be initiated by bacteria, the yeasts may be introduced by cactophilic *Drosophila* that feed and breed in the rotting tissue.

Plant Pathogenic Yeasts

Some yeasts are recognized as plant pathogens, but their pathogenicity is of a mild form compared with that of phytopathogenic bacteria and higher fungi. (For the purpose of this discussion we exclude yeasts that cause an active fermentative spoilage of soft, ripe fruits, since the yeast action is as a rule of a secondary nature after the fruit has been damaged in some way.)

Yeasts responsible for plant diseases are primarily members of the genera *Nematospora* and *Ashbya*, the latter a yeast-like organism. Species of both genera have characteristic needle-shaped ascospores. The two best recognized species, *Nematospora coryli* and *Ashbya gossypii*, have been reported in a number of host plants, almost exclusively from subtropical and tropical areas. Typical examples of host plants include cotton (the internal boll disease), lima beans and other legumes (yeast spot), coffee berries (coffee bean disease), tomatoes, pecans, hazelnuts, and citrus fruit. The condition in citrus fruit is called "inspissosis" and causes a local drying out and collapse of the juice sacs and a wrinkling of the rind due to the growth of the yeast. The cotton boll disease is of considerable economic importance. The growth of the yeasts in immature, unopened bolls leads to a staining of the lint. The evidence points to hemipterous insects—bugs of the genus *Dysdercus* in particular—as the carriers or vectors of the disease. Spores or cells of *Nematospora* are carried by *Dysdercus* species on the mandibles or in stylet pouches. They enter the plant tissue during feeding by the insects. A fungal disease transmitted in this way is referred to as "stigmatomycosis."

Yeasts from fruits and berries are discussed in Chapter 9.

YEASTS ASSOCIATED WITH ANIMALS

Nearly all of the information available on yeast–animal relationships pertains to various warm-blooded species (chiefly herbivorous mammals), insects, and crustaceans.

Warm-blooded Animals

Differentiation should be made between yeasts that form part of the normal intestinal flora and those responsible for certain pathological conditions. In surveying the yeasts occurring in the alimentary tract, investigators have derived information from fecal matter as well as direct sampling of the intestinal or stomach contents. These studies have shown that about half of the samples obtained from horses and cattle contained yeast. Swine have a high incidence of yeasts (close to 90%), which is due to the very regular occurrence of *Candida slooffii* in these animals. Similarly, rabbits have a very high incidence of yeasts, nearly always represented by a single species, *Cyniclomyces guttulatus*. A number of small rodents, such as mice and rats, regularly carry *Torulopsis pintolopesii* and *T. bovina* (the latter having a sexual stage, *Saccharomyces telluris*). On the other hand, sheep and goats usually have a very low incidence of yeasts in their intestinal tract.

All of the yeasts mentioned here might be considered as obligate parasites of their particular hosts, since they grow poorly or not at all at room temperature or below, they have unusually stringent growth requirements, and have not (or very rarely) been isolated from natural substrates outside of the host. These organisms are not known to benefit their hosts. A number of other species may be considered as facultative parasites of warm-blooded animals—that is, yeasts that are regularly isolated from sources outside the animal body, but which, due to their ability to grow at 37°C, can also multiply in the intestinal tract. Examples of such yeasts are *Candida krusei, C. tropicalis, C. parapsilosis,* and *Trichosporon cutaneum.* These yeasts and other normally saprophytic species can become opportunists when patients are in a debilitated condition or are treated with broad spectrum antibiotics that destroy the bacterial flora.

Finally, a large number of miscellaneous yeasts have been isolated sporadically from intestinal contents or dung that may be considered as incidental transients, having been ingested with a particular food. Many of these species cannot grow at 37°C and may even be killed at that temperature. However, even if a yeast has the ability to grow at body temperatures, it may be unable to grow in the intestinal tract because of anaerobiosis or pH conditions.

Nonpathogenic yeasts also occur on the skin of animals and humans. Such organisms usually originate from various external sources with which the skin may come into contact. Species of *Pityrosporum* deserve special mention because of their frequent occurrence on the scalp of humans and on the skin of animals. *Pityrosporum* is often associated with dandruff, but not necessarily the cause of it. *P. ovale* (from human scalp) requires lipids for growth (see Chapter 7). And *P. pachydermatis* (from the ear wax of dogs) can grow without fat but is stimulated by its presence. *Pityrosporum* is the yeast (culture) phase of the dermatophytic fungus *Malassezia.*

In the vast field of medical mycology yeasts play a relatively minor role, although in some instances yeast infections have caused serious illness and even death.

Our discussion is limited to the true yeasts, but note that a number of medical fungi assume a yeast phase in the body of the host, whereas in culture they are typical fungi—for example, *Histoplasma capsulatum* (ascogenous stage *Emmonsiella capsulata*, the cause of histoplasmosis), *Blastomyces dermatitidis* (the cause of North American blastomycosis), and *Coccidioides immitis* (the cause of San Joaquin Valley fever, or coccidioidomycosis).

Yeasts that can, under certain circumstances, be responsible for infections include *Candida albicans, C. stellatoidea, C. tropicalis, C. parapsilosis, Torulopsis glabrata,* and *Cryptococcus neoformans*. In general, these yeasts must be considered as opportunists; they may cause pathologic conditions in susceptible debilitated individuals—often those in poor nutritional status or those suffering from diabetes. *Candida albicans*, which causes various forms of candidiasis or moniliasis of the skin, and particularly of mucous membranes (for examples, thrush and vaginitis), is quantitatively the most common pathogenic yeast. Surveys of many animals have shown, however, that man, fowl, and hedgehogs are very common carriers of this yeast, apparently without ill effects. Only in cases of diabetes in man, debilitating disease, pregnancy, and especially upon treatment of patients with broad spectrum antibiotics (when the balancing bacterial flora is destroyed) is the disease likely to become apparent and even systemic. Other species of *Candida* (for example, *C. parapsilosis*) are encountered more rarely as aetiologic agents for disease, although they may accompany infections by certain fungi. Several reports in the literature indicate that *Candida parapsilosis* may be responsbile for heart valve infection in heroin addicts, often with fatal results. *Torulopsis glabrata* has been isolated with some degree of frequency from infections of the urinary tract.

Cryptococcosis (torulosis), a disease caused by *Cryptococcus neoformans*, occurs relatively rarely, but proved cases are often dramatic and fatal. At first entailing skin lesions, the yeast infection becomes systemic and involves the central nervous system, often causing a chronic meningitis ending in death. The most important reservoir of this yeast in the animal world is believed to be pigeons, in whose droppings and nests the yeast has been repeatedly demonstrated by Emmons and others. The sexual form of this yeast is *Filobasidiella*, a basidiomycetous genus.

Insects

Since insects are probably the most important vectors in the distribution of yeasts in nature, their relationship to yeast is particularly important. The frequent association of yeasts with insects is undoubtedly a result of the nutritional importance of microorganisms in the life cycle of many species of this very large class of animals.

A good illustration is found in members of the genus

Drosophila. Many of its 1,200 or more described species feed on a variety of substrates that contain yeasts able to carry out an alcoholic fermentation. There is some evidence that fermentative yeasts are associated with adults, while oxidative species are more commonly found in larval habitats, such as slime fluxes. A number of extensive surveys have shown the regular presence of yeasts in the crop or in the intestinal tract of both the domestic fruit fly *D. melanogaster* and in a number of wild species. As was pointed out earlier in this chapter, the yeasts are actually digested; for this reason the number of yeast cells per fly depends on the available food supply and on the time between the last feeding and dissection of the insect.

The feeding of *Drosophila* on moist fermenting fruit is well known, and breeding by *D. melanogaster* commonly occurs in the same substrate. Thus, in fruit orchards, in tomato fields, and near dumps where fruit residue is discarded, the food supply is not limiting. Consequently, this species may develop into astronomical numbers. The most common yeasts found in the crops of *D. melanogaster* are the apiculate yeasts, *Hanseniaspora* and *Kloeckera*, and also *Pichia kluyveri*, *P. fermentans*, *Candida krusei*, and *Torulopsis stellata*.

The situation is quite different with wild species, which apparently are not tempted to utilize food that is easily but sporadically available. Their populations fluctuate much less, and their biomass in a given area must be a function of the availability of a steady food supply. The yeasts that are common in the crops of wild species of *Drosophila* in central California are *Saccharomyces montanus*, *S. cerevisiae* var. *tetrasporus*, *Hansenula polymorpha*, *Kluyveromyces veronae*, and *Kluyveromyces drosophilarum*. In spite of an extensive search for substrates containing these yeasts, the results have been largely negative. Some of the areas studied are extremely arid, where substrates with enough surface moisture to permit feeding are scarce or small and ephemeral. One source of yeasts in the forests of the Sierra Nevada of California (where most of our studies were conducted) is exudates of oaks and firs. Apparently, this source is not attractive as a source of food for adult flies, although some species use these slime fluxes for breeding purposes and larval development. The fundamentally different yeast florae in the crops of adult flies and these fluxes is the basis of this conclusion.

There are still other sources of yeasts that again are very different in composition from the flora found in adult drosophila flies and in slime fluxes. These yeasts are abundantly associated with bark beetles that attack coniferous trees.

The two genera of bark beetles that commonly infest pine trees are *Ips* and *Dendroctonus*, whereas beetles of the genus *Scolytus* more commonly attack true firs (*Abies*) and Douglas firs (*Pseudotsuga*). The larval galleries of these insects are made in the phloem of the tree and result in rapid death of the host. A new crop of adults originating from these larvae then emerges through the bark and, after dispersal, selects another susceptible tree. After boring through the bark they oviposit and the cycle starts anew. Yeasts are found in large numbers in the galleries made by larvae and adults, in spite of the presence of oleoresins in the wood. Perhaps these chemicals have a selective action, since the yeast flora is very specific and consists mainly of weakly fermentative or oxidative types. *Hansenula capsulata, H. wingei*, and *Candida silvicola* (the perfect stage of which is *Hansenula holstii*) are abundant in all species of pine trees; *Pichia pini* occurs in all except *Pinus jeffreyi*. Possibly the high concentration of n–heptane in the turpentine of this pine might have a selective inhibitory action on *Pichia pini*. Other pines lack this constituent.

In contrast, *Scolytus* beetles in firs and Douglas firs usually contain nearly pure cultures of an entirely different yeast, *Pichia scolyti*. Although this heterothallic yeast is occasionally found in association with *Ips* and *Dendroctonus* in pines, species of *Scolytus* must be considered as its true vectors.

Several species of *Saccharomycopsis* are associated with ambrosia beetles in deciduous trees. Again, none of these yeast species has been isolated from adult drosophila flies.

Although it is still not known where many of the wild, adult drosophila flies (excepting cactophilic desert species) obtain their food in nature, these small insects nevertheless are extremely useful in surveying (at least in part) the yeast flora in unexplored areas. Drosophila flies have a worldwide distribution. Techniques have been developed to capture these insects after attracting them to a sterilized or screen-covered bait and to plate out the contents of their crops for the yeasts present.

Another illustration of a characteristic association between an insect and a yeast is one involving the tiny fig wasp *Blastophaga psenes*. A brief explanation of its life cycle is necessary. The fig is a syconium—that is, a hollow receptacle lined with flowers. It contains an opening, or eye, essentially covered by scales, through which certain small insects may pass. The normal habitat of the fig wasp is the caprifig, an inedible variety, which produces three successive crops each year. Only the spring crop (profichi) produces both staminate and pistillate flowers. The fig wasp, which lives inside the fig, carries pollen from the profichi crop to the subsequent crop so that seeds are formed. The insect passes from one crop to the next until the cycle repeats itself. We have found an interesting microflora, quite constantly associated with the fig wasp, consisting of one species of bacterium, *Serratia plymuthica*, and a single species of yeast, which has been named *Candida guilliermondii* var. *carpophila*. Presumably these two microorganisms are useful to the fig wasp, either nutritionally or otherwise. We would digress too far to discuss the fascinating life cycle of the fig wasp in greater detail and to point out how it is used in the commercial production of edible figs. For further information refer to Condit's book, *The Fig*.

Recently, investigators in the Moscow region of the USSR found that two species of *Debaryomyces*, *D. for-*

micarius and *D. cantarellii*, are present in large numbers in ant hills of the *Formica rufa* group.

Several highly specific associations exist between yeasts and certain specialized tissues in the intestinal tract of insects, where these yeasts have established themselves as harmless intracellular parasites. Some of these yeasts have been obtained in pure culture, but others have been seen only in their natural habitat and have resisted efforts to isolate them.

The few examples cited, in relation to the enormous number of insect genera known, leads one to believe that the study of yeast–insect relationships has barely been initiated. The interesting specific associations so far discovered promise many additional fascinating habitats in other groups of insects.

Crustaceans

Outside the areas of warm-blooded animals and insects, relatively little work has been done with regard to yeast habitats. This again points to fruitful areas of research, both from the standpoint of yeast ecology and from the point of view of the related animal species. The finding of *Metschnikowia bicuspidata* as a pathogen of *Daphnia magna* (a fresh water crustacean) and *Artemia salina* (the brine shrimp) points in this direction. Unidentified species of *Metschnikowia* have also been implicated in diseases of copepods during certain seasons of the year. Interestingly, the terrestrial species of this genus are usually isolated from fruits and from flower nectar. These species are probably carried by bees.

YEASTS IN SOILS

In most cases, the soil should be considered more as a reservoir than as a habitat where yeasts can multiply freely. Since the soil receives plant, fungal (mushroom), and animal residues of all sorts, one might expect to isolate a large variety of yeasts from soil, which is true within limits. The yeast flora is controlled, however, by such factors as (1) the balance of death and growth rates; (2) longevity of a particular species, coupled with the ability to withstand local competitive soil organisms (fungi, bacteria, protozoa, and nematodes), especially those producing antibiotics; (3) types of higher plants growing in the particular soil and the kinds of fruits or seeds they produce, as well as leaves and exudates that fall on the soil; (4) the types of mushrooms and other fleshy fungi that grow on a particular soil; (5) animals living and dying on a particular soil; and (6) soil composition, season, climate and temperature, sun exposure, depth under the surface, moisture content, and so on.

A great variety of yeasts have been isolated from soils by a number of investigators, starting with the early work by Emil Christian Hansen, and more recently by Lund, di Menna, and Capriotti. Yeasts are found in tropical soils as well as in those from arctic or antarctic regions. Only the more recent surveys have some quantitative value, since nearly all of the earlier yeast isolations were made by enrichment techniques. Thus, it is not surprising that highly variable results have been obtained. The total yeast count is usually rather low, as compared with the numbers of bacteria and fungi present. Populations range from none or a few cells to several thousand per gram of soil from fields, meadows, gardens, forests, peat bogs, etc. Only in certain soils taken under berry bushes, grape vines, and fruit trees, which may deposit spoiled fruits with large numbers of yeast, have higher counts (up to 250,000 cells per gram) been recorded.

The yeast population is also affected by its depth in the soil. Yeasts are most numerous in the upper layers, from approximately 2 to 10 cm in depth. In the deeper layers they become progressively scarcer and are exceedingly rare at a depth of 30 cm. Samples from the very top have very low counts if the surface is subject to intense sunlight, heating, and desiccation. The vertical distribution of yeasts in soil depends on such factors as compaction and porosity, rainfall, cultivation, burrowing animals, and the presence and movement of soil-inhabiting insects that feed and breed on decomposing fruits (for example, the dried-fruit beetle, *Carpophilus hemipterus*).

Consideration should also be given to the production of yeast nutrients by other soil microorganisms. For example, bacteria and fungi can degrade substrates, such as cellulose and certain other polysaccharides, that are not utilizable by yeasts until they are depolymerized. This could explain why, in an experiment by Lund, soil yeasts inoculated into a sterilized soil had essentially died after six months, but in a control experiment with unsterilized soil the natural yeast flora had increased greatly in the same period. A factor that has not been studied in any detail, but which undoubtedly plays a role, is the consumption of yeast by small soil-inhabiting animals, such as nematodes. Replenishment of the soil flora with yeasts from plants and animals can be a selective process. For example, di Menna has clearly shown that the yeast flora found on pasture plants was different in several respects from that of the soil below—even as shallow as 1 inch below the surface.

We cannot list all of the yeasts that have been isolated from soil. However, thus far species of some genera have been isolated exclusively from this source, although it could be that their true habitat is elsewhere. Examples are the species of *Lipomyces*, *Schwanniomyces*, and *Schizoblastosporion*. Examples of members of other genera that have been obtained frequently or exclusively from soil are *Hansenula saturnus*, *H. californica*, *H. suaveolens*, *Debaryomyces castellii*, *Candida humicola*, *C. scottii* (a psychrophilic yeast, whose perfect form belongs to *Leucosporidium*), *Cryptococcus terreus*, and certain other species of the last genus.

It is of considerable interest to know if potentially pathogenic yeasts are free-living and whether they occur in soil. Such occurrence is well known for pathogenic fungi—for example, *Coccidioides immitis*, the cause of coccidioidomycosis or valley fever. Both *Candida albicans* and *Cryptococcus* (perfect *Filobasidiella*) *neoformans* have been found, the former by direct plating, but the latter only after enrichment by injecting a soil suspension intraperitoneally into mice. Neither yeast is very common in soil, and they are probably introduced via the feces of various wild and domestic animals (for example, *Cryptococcus neoformans* in avian droppings).

YEASTS OCCURRING IN WATER

Yeasts are found in variable numbers both in fresh and in salt water. Our knowledge of aquatic yeasts (especially those in fresh water) is much more limited than that of the terrestrial yeasts. From what we have discussed in the previous sections, and in particular that pertaining to soil yeasts, waters obviously obtain yeasts from many sources. Considerable terrestrial contamination can be expected in rivers, lakes, and coastal marine regions, and positive evidence in this direction has been obtained. On the other hand, yeasts can also propagate on the aquatic fauna and flora characteristic of the environment—for example, algae, marine grasses, trees, and even plankton of the open oceans.

Yeasts from Salt Water

Surveys have shown that yeast populations are most dense in coastal waters, but they are also found in mid-ocean and at depths of 4,000 meters, although in small numbers. Population densities are undoubtedly contingent upon the availability of organic substrates for growth. Sea water normally contains between 10 and 100 yeasts per liter. But close to grass and algal beds, where decomposition is going on, the number of viable yeasts may be 5,000 to 6,000 per liter. On decomposing plankton and seaweeds the counts are much higher. The yeasts encountered in ocean water are generally asporogenous, oxidative types that are not fundamentally different from the same or related species isolated from terrestrial sources. Examples are *Candida diddensii*, *C. polymorpha*, *Rhodotorula rubra*, *Rhodotorula marina*, and *Cryptococcus laurentii*. Among the sporogenous yeasts are *Debaryomyces hansenii*, *Rhodosporidium sphaerocarpum*, *Rhodosporidium dacryoidum*, *Leucosporidium scottii* (from antarctic seawater), and several species of the genus *Metschnikowia*. *M. zobellii* and *M. krissii* were obtained for the first time in pure culture by van Uden in 1961 from marine sources off the Southern California coast. More recently *M. bicuspidata* var. *bicuspidata* was obtained from brine shrimp collected in a Canadian salt lake. Isolation of these species is of particular interest, since a yeast with needle-shaped spores was described as early as 1884 by the Russian microbiologist Metschnikoff in its host *Daphnia magna*, but he could not isolate it in culture.

Sympodiomyces and *Sterigmatomyces*, two recently described genera, have been isolated from ocean waters. *Pichia spartinae* is very common in decomposing oyster grass *(Spartina alterniflora)* in marine swamplands of southern Louisiana, while *Kluyveromyces aestuarii* has been isolated from estuarine waters in Florida.

Yeasts from Fresh Water

The knowledge of yeasts occurring in fresh water is more limited, but the available data again point to weakly fermentative or oxidative species. Examples of yeasts that according to published reports have been isolated repeatedly from lakes and rivers are *Cryptococcus laurentii, Torulopsis candida, Candida krusei,* and species of *Sporobolomyces.* The yeast flora of such bodies of water can be strongly influenced by pollution from terrestrial sources, including sewage.

FUTURE RESEARCH IN YEAST ECOLOGY

In spite of a large number of reports on the distribution of yeasts in nature, many existing areas of research could be fruitfully expanded in new geographic regions. Hitherto unexplored substrates and niches could reveal additional habitats and associations for yeasts. For example, a continuation of studies on the association of yeasts with *Drosophila* species, and discovery of their feeding and breeding sites in unexplored areas of the world should prove highly rewarding on the basis of past results. The same applies to bark and ambrosia beetles in various tree species, as well as other insects depending on yeasts as part of their diet. Studies of yeasts associated with plants should be extended to the little explored groups of ferns, mosses, algae, and fleshy fungi; the few studies made on the phyllosphere of higher plants should also be expanded. A fascinating area may prove to be a study of mites in yeast dispersal. As is well known in many laboratories, these small arachnids (order *Acarina)* often invade petri plates in search of yeasts. Some mites are ectoparasites of insects (for example, *Drosophila).* They can be carried by such flies from one substrate to the next, thus becoming responsible for the dispersal of yeasts upon which drosophila flies feed. Some mites have piercing mandibles; these may be responsible for initiating bacterial and yeast infections in plant tissues, such as necrosis in the stems of cacti. Other areas of study worth expanding are concerned with airborne yeasts and their air dispersal (including the effect of high altitudes), and with the presence and origin of yeasts in fresh water lakes, especially those free of pollution. We are convinced that imaginative research by yeast ecologists will lead to the discovery of many additional interesting associations and the isolation of novel yeast species.

9 / Yeast Spoilage of Foods and Fermentation Processes

Spoilage yeasts are defined as organisms that produce undesirable changes in foods or during fermentation processes. These changes may be limited to no more than an aesthetic alteration of the product by the physical presence of yeast. This may appear in the form of a pellicle or a turbidity in liquids, or as a slimy or powdery coating on solid products. In some instances substantial yeast growth may cause undesirable changes due to metabolic products of yeast, such as the formation of unnatural odors or flavors, or to metabolic activity causing an increase in pH due to utilization of organic acids. For example, lactic, citric, or acetic acids are used extensively as food preservatives, and their removal or reduction in concentration by yeasts can encourage spoilage bacteria to develop.

Even benzoic, propionic, and sorbic acids, sometimes used as preservatives in certain acidic food products, can be utilized by a limited number of yeast species. For example, several species of *Rhodotorula* grow well on 0.25% benzoic acid as a carbon source (pH 4.5); *Saccharomyces rosei* can grow on 0.25% propionate (pH 4.5), and *Brettanomyces intermedius* grows in the presence of 0.1% sorbic acid (pH 4.8). Utilization of sorbic acid is generally slow and poor by most yeasts. Fortunately, the presence of spoilage yeasts in food has never resulted in food poisoning phenomena. The metabolic products of yeast are not considered toxic, and the yeasts themselves, even though some pathogenic species exist, are not known to be responsible for food-borne infections or poisonings, as is the case with a number of bacterial and fungal species.

Yeasts responsible for food spoilage are often well-known species. These "free-living organisms" begin multiplying when the opportunity presents itself. The composition of a particular food strongly influences which species are favored in their development. From this point of view, therefore, spoilage of foods constitutes a special aspect of ecology. The selective conditions that are established by the chemical composition of the product, the manner in which it is packed, the oxygen concentration in the gas phase of the container, the temperature, and other factors of storage usually limit the number of species to relatively few major ones. Even temperatures of 0–4°C, which are slightly lower than those found in most household refrigerators, do not protect products from growth and incipient fermentation by yeast, provided the storage period is several weeks or longer.

The sources of the yeasts that initiate spoilage of foods vary greatly with the product. Sound fruits and animal tissues are inherently sterile, but may become contaminated during harvest, transport, and in processing plants from improperly cleaned equipment. Typical examples are wooden vats and cutting tables, pumps, pipelines, and other processing equipment. In addition, the outer surface of fruits carries yeasts that were probably deposited by insects, dust, and wind. A few spoiled fruits in an otherwise sound lot can increase the microbial contamination tremendously. Finally, yeasts may enter by way of added ingredients, such as salt, condiments, and spices. Several yeast cells can develop into astronomical proportions given enough time and the proper selective environmental conditions.

It is most convenient to discuss the spoilage of foods from the standpoint of broad categories in composition rather than to list particular spoilage yeasts that have been isolated from individual products at various times. In doing so, we illustrate typical causative organisms for particular kinds of products.

YEASTS OCCURRING IN SUBSTRATES WITH A HIGH SUGAR CONTENT

Such products include honey, jams, jellies, syrups, dried fruits, fruit juice concentrates, molasses, and other products preserved by high sugar concentrations. Yeasts that are able to grow in such an environment are usually referred to as "osmophilic" yeasts (yeasts that grow best in an environment of high osmotic pressure) or osmoduric yeasts (yeasts that can tolerate high sugar concentrations). We are speaking of sugar concentrations of 40–70% by weight. Yeasts are not easily inhibited by moderate sugar concentrations; at 40% sugar (which may be encountered in homemade jams or canned fruit), a large percentage of the known species will grow well, and selectivity of the more osmoduric types occurs above 40%. In general, the higher the sugar concentration, the slower the rate of growth. There are few organisms that can grow at sugar concentrations between 65–70%. Hence, spoilage in products of very high sugar content may become apparent only after many months of incubation or storage. However, several species can grow moderately well at 60%, and significantly more will grow when the sugar concentration is lowered to 50%. Hygroscopicity is a factor that may favor the development of yeasts as spoilage organisms in solutions of high sugar content stored in atmospheres of high relative humidity. This results in a surface layer slightly more dilute than the rest of the sugar solution, and yeast growth may well start in such thin films at the surface of the liquid.

Examples of yeasts that are recognized for their osmophilic properties are several haploid species of the genus *Saccharomyces,* in particular *Saccharomyces rouxii, Saccharomyces bisporus* var. *mellis* and *Saccharomyces bailii* var. *osmophilus.* These species are most frequently responsible for spoilage in such substrates. Some strains of *Saccharomyces rosei*, a common species, are also recognized for their high osmotolerance. A less common organism is *Schizosaccharomyces octosporus*, which was first found by Beijerinck in 1894 on spoiled currants from southern Europe. We have isolated this yeast repeatedly from the sugary coating that sometimes develops on the surface of dried figs, dates, and prunes. *Eremascus albus* is an obligately osmophilic yeast-like organism that occasionally is isolated from concentrated products (for example, malt and dried mustard). Other examples are the advanced, free-living species of *Hansenula*, such as *H. anomala* and *H. subpelliculosa*, and the species of *Hanseniaspora*—one of which was appropriately named *Hanseniaspora osmophila*. Foods that have undergone spoilage by osmophilic yeasts are usually characterized by a slight odor of fermentation caused by esters, aldehydes, and other volatile products of yeast metabolism. The yeast concentration is fairly low as a rule.

YEASTS OCCURRING IN PRODUCTS OF HIGH SALT CONTENT

The osmotic pressure caused by high concentrations of sodium chloride generally does not favor the development of the same yeasts as are found in substrates of higher sugar content. One exception to this rule is *Hansenula subpelliculosa*, which occurs in both environments. The development of yeasts in products of high salt content is characteristically associated with the production of fermented olives, cucumbers (pickles), cabbage (sauerkraut), etc. These products undergo a lactic acid fermentation after being covered with a brine solution. In the case of olives, a final concentration of about 6.5% salt in the brine is typical. But for cucumbers the salt concentration is considerably higher; depending on the process, it ranges from 10–16%. During the lactic acid fermentation, which takes place at the expense of the natural sugars in the fruit, the pH drops. Simultaneously a characteristic fermentative yeast flora develops in the liquid. These subsurface fermentative yeasts consist of several species of *Torulopsis* (for example, *T. etchellsii*, *T. versatilis*, and *T. holmii*), some species of *Saccharomyces*, and *Hansenula subpelliculosa*.

When the sugar is used up and the pH has dropped because of lactic acid production by the lactic acid bacteria, a secondary, oxidative yeast flora develops on the surface of the liquid in the form of a thick, folded layer of yeast. This secondary surface flora consists of species of *Pichia* (for example, *P. membranaefaciens*) in brines of lower salt concentrations. Also , the imperfect yeasts *Candida krusei* and *C. valida* are common. With higher salt concentrations, species of *Debaryomyces* become more prevalent. In fact, species of *Debaryomyces* (in particular *D. hansenii*) are known as the most salt-tolerant yeasts, being able to grow in nearly saturated brine solutions. Since *Debaryomyces hansenii* is a common yeast in ocean waters, it could survive during the production of solar salt, and in this way arrive in brined products. In meat brines, in salted meat products, and on the surface of certain cheeses, the most common yeasts constitute species of this genus. The slimy films that occasionally develop during cold storage on bacon, ham, wiener sausages, and pickled meats often contain *Debaryomyces* in large numbers. In this connection natural sausage casing in the form of animal gut is frequently preserved with crude rock salt; in these salted casings, we have also found *Debaryomyces* species in abundance.

YEASTS ASSOCIATED WITH PRODUCTS THAT HAVE A MODERATE SUGAR CONTENT

Examples of such products are fresh fruits and unpasteurized fruit juices. In most of these the sugar content in percentage by weight ranges from about 8–15% or for grapes to 24% or more. Many groups of investigators have studied the yeast flora of fermenting fresh fruits or fruit juices. Many yeast species have been found. Characteristic examples are *Torulopsis stellata, T. cantarellii, Pichia kluyveri, P. fermentans,* and species of *Hanseniaspora* (including its imperfect genus *Kloeckera*), which are extremely prevalent in such products. In this connection the very early stages of natural wine fermentations (in which sulfur dioxide and pure culture inocula are not used) always show the presence of large numbers of these apiculate yeasts in addition to smaller numbers of *Saccharomyces, (for example, S. bayanus, S. capensis,* and *S. rosei), Torulopsis,* and *Candida* species. *Saccharomyces cerevisiae* (wine yeast) is a minority organism during the early stages of natural wine fermentation. In our experience typical strains of *S. cerevisiae* (excluding *S. cerevisiae* var. *tetrasporus*) are rarely if ever present on the fruits and berries of wild species of plants.

In recent years a number of reports have appeared in which species of *Brettanomyces* were shown to be responsible for developing turbidity in bottled, carbonated soft drinks. *Brettanomyces abstinens* and *B. naardenensis* are two species recovered repeatedly from ginger ale and from lemonade or tonic water, respectively. The pH of the latter beverages ranged from 2.6–3.2.

YEASTS ASSOCIATED WITH MILK PRODUCTS

Since the principal sugar of milk is lactose, many of the spoilage yeasts in dairy products are lactose fermenters, or at least yeasts that can use the sugar lactose as an energy source (for example, *Kluyveromyces lactis*, *K. fragilis*, and several lactose-positive species of *Candida* and *Torulopsis* are commonly found in spoiled milk products). Because condensed milk is sweetened with sucrose (final concentration approximately 40–45%), non–lactose-fermenting species of yeasts may be responsible for a gassy swelling of cans containing this product. Two species responsible for this defect are *Torulopsis lactis-condensi* and *T. globosa* (the latter shown to be the imperfect form of *Citeromyces matritensis*), both of which are tolerant to high concentrations of sugar.

"Pink" yeasts of the genus *Rhodotorula* are frequently present in milk and cream. Sour milk and cream sometimes develop red spots on the surface due to the growth of these aerobic, lactic acid-tolerant species of yeast.

SALAD DRESSING

Occasionally, uncommon yeast spoilage processes have come to our attention. We handled an unusual case where a species of *Torulopsis* caused a gaseous spoilage in salad dressing. This mayonnaise-like salad dressing normally contains fermentable carbohydrates (in contrast to true mayonnaise) and is therefore more susceptible to spoilage. It has a high content of acetic acid, however, and the yeast we isolated had an unusual tolerance to this acid. In pure systems this yeast was able to grow in nutrient solutions in which the pH was lowered with acetic acid to 2.5. After about two weeks, evident gaseous fermentation had taken place, and this yeast caused active spoilage upon reinoculation of the commercial product. The growth of most yeasts is already inhibited by acetic acid at a pH of 3.5.

FATS AND OILS

The spoilage of pure fats and oils by yeast is rare, but products such as butter and margarine, which contain an aqueous phase in addition to a fat phase, are sometimes subject to yeast spoilage. Such yeasts—for example, *Candida lipolytica* (ascogenous form *Saccharomycopsis lipolytica*) and C. *steatolytica*—possess potent lipases and can use fats as energy sources after hydrolyzing them to glycerol and fatty acids. *Candida lipolytica* has also been isolated from olives,—another product with a high content of oil in an aqueous phase.

STARCHY PRODUCTS

Several yeasts have the capacity of growing on soluble starch as carbon source by virtue of their ability to produce various extracellular amylases. These include α– and β– amylase, as well as glucamylase. Usually a mixture of these enzymes is excreted by yeasts growing on starch. Some of these species grow on starchy food and cause spoilage. *Saccharomycopsis fibuligera* is an example of a species that has been isolated several times from "chalky bread," in which the bread is covered on the surface with dull white areas consisting of masses of yeast cells. Other products that are susceptible to this type of spoilage under conditions of high humidity are macaroni, flour, rice, other cereals, and cereal products. Other species of *Saccharomycopsis* with strong amylolytic activity are *S. capsularis* and *S. malanga.* All species of *Schwanniomyces* grow well on soluble starch. However, these last organisms, which have been isolated only from various soils, have not been implicated thus far in food spoilage. Some species of *Hansenula* can utilize starch, and these species are sometimes associated (although not necessarily as spoilage organisms) with oriental fermentation processes of rice.

WINES AND BEER

Although table wines have very little residual sugar, development of "wild" yeasts (as contrasted to the original wine yeast) in bottled wines, resulting in turbidity and sometimes off-flavors, is known to occur periodically. From bottled table wines (including white and red wines as well as champagne) we have isolated *Saccharomyces bayanus, S. capensis, Brettanomyces intermedius,* and *Pichia membranaefaciens. Brettanomyces intermedius* and its ascogenous form *Dekkera intermedia* also have been isolated with increasing frequency from South African and European table wines. One reason for the better recognition of these last two species is the use of a selective medium, containing 100 mg/l of the antibiotic cycloheximide (to which these yeasts are resistant, in contrast to wine yeasts and most other common spoilage yeasts) and an increased level of thiamine (10 mg/l) that these yeasts require for adequate growth. Even so, colonies of *Brettanomyces* and *Dekkera* are slow growing. Plates should be kept for at least one week before drawing conclusions about their absence. Very rarely is turbidity due to yeast growth encountered in fortified sweet wines of 18–20% alcohol, but yeast development is somewhat more common in sweet sauterne wine of 14% alcohol and low sulfur dioxide content. One organism involved in such spoilage is *Saccharomyces bisporus* var. *mellis,* a strongly osmoduric species. In these environments the growth is slow and limited, but it has a detrimental effect on the appearance of the wine.

Although in the past certain species of *Brettanomyces* were used for the production of special kinds of ale in Ireland and Belgium, species of that genus as well as its ascogenous form *Dekkera* are now generally considered as spoilage yeasts of unpasteurized beer. A number of reports have appeared in the literature describing the development of such yeasts in beer and the resulting turbidity and off-flavors. In the presence of sufficient oxygen these yeasts oxidize alcohol to acetic acid. Under anaerobic conditions they appear to grow fermentatively on residual sugar and dextrins.

Other yeasts (species of *Saccharomyces*) are able to form higher alcohols in beer from certain amino acids by the Ehrlich process (see Chapter 6). Tyrosol, formed from the amino acid tyrosine, has an intensely bitter flavor, whereas 2-phenylethanol, formed from the amino acid phenylalanine, has an undesirable rose-like aroma.

An example of a yeast that produces specifically glucamylase is *Saccharomyces diastaticus*. This enzyme removes single glucose residues from the nonreducing termini of the outer chains in starch and dextrins. It can also hydrolyze the α–(1→6)-linked branch points so that potentially the entire substrate molecule may be degraded. This yeast has been identified as a spoilage organism of unpasteurized beer. Since it can hydrolyze residual dextrins in beer to glucose and ferment the sugar thus formed, it causes the phenomenon of superattenuation, giving rise to excessive CO_2 pressure and too much alcohol. Such beers, lacking extract, are inclined to suffer from "gushing" or excessive foaming.

BAKER'S YEAST

The occurrence of "wild" yeasts with baker's yeast may lead to the more rapid multiplication of these contaminants during the production stages and to a decrease in the baking quality of this highly standardized product. We have had opportunity to observe the competitive behavior of *Candida utilis* (a well-known feed yeast) in a baker's yeast plant where it had established itself as a contaminant. Since *Candida utilis* grows faster than baker's yeast, it occasionally amounted to more than 10% of the final yeast crop in the end product—in spite of a very minor initial contamination. This particular contamination and competition could be conveniently studied by the property of *Candida utilis* to utilize nitrate as a single nitrogen source, which *Saccharomyces cerevisiae* cannot do. On differential media we could follow the development of the contaminant in competition with baker's yeast.

In storage, cakes of compressed baker's yeast sometimes become covered with a chalky white coating consisting of an aerobic, yeast-like fungus *Geotrichum candidum*. This organism, which reproduces vegetatively by fission and arthrospores, may impart an off-flavor to the yeast and it also detracts from the appearance of the yeast cake.

Geotrichum candidum grows on many food products besides compressed baker's yeast cakes. Economically, it is better known as the cause of "sour rot" on citrus fruits (particularly lemons), soft rot of tomatoes, and as a common postharvest rot of various fruits and vegetables. It is commonly isolated from soil samples, and the food regulatory agencies have made use of it as an index of plant sanitation. Its habit of growing in environments of high humidity, such as on moist surfaces of food plant machinery (for example, conveyors and belts), as a white mycelial mass has given it the name of "machinery mold."

Although many interesting cases of food spoilage by yeasts have been encountered and would be worthy of a more detailed discussion, the examples mentioned show the diversity of yeasts as far as food spoilage organisms are concerned. Almost any kind of food will permit yeasts to grow if it has not been adequately heat treated. High concentrations of sugar, salt, organic acids, the exclusion of air, refrigeration, and application of other storage conditions will not safeguard a food from the action of yeasts, provided storage is sufficiently long. The heavier the initial contamination, the sooner spoilage symptoms become apparent. For this reason strict observance of sanitation in food-processing plants offers the best protection against losses caused by microbial spoilage.

10 / Industrial Uses of Yeast

Yeasts and their metabolic products are also utilized by man. As was explained in Chapter 1, some of the activities of yeasts have been used on an empirical basis for many centuries; today, however, yeasts are employed in many ways, usually in well-equipped plants and under proper scientific control.

Fermented Beverages

Two broad classes of raw materials serve as substrates for the production of fermented beverages. The first type of substrate contains a supply of directly fermentable sugars (principally fruits and berries), and the second a potential source of fermentable sugars, mainly in the form of starches (for example, cereal grains). Since most industrial yeasts cannot ferment starch and other polysaccharides, various methods have been developed for converting these polysaccharides into fermentable sugars.

Beer Brewing

The primary raw material in the brewing process is barley malt. Essentially, malt is made from special varieties of barley, that are allowed to undergo a limited germination process in specialized plants called "malt houses". During this germination period, lasting from five to seven days, the necessary starch-splitting enzymes and various other enzymes are formed in the grain. Although the enzymes penetrate into the starch-containing endosperm of the kernels, starch hydrolysis during malting is limited to a small amount of β–amylolysis—that is, the removal of some maltose residues from starch by β–amylase. However, amino acids and peptides are formed by proteolysis during this process. The germinated barley or malt is then dried to a low moisture content at carefully controlled temperatures to preserve the enzymatic activity.

In the brewery the malt is crushed to a flour and exposed to warm water at approximately 60–70°C, during which starches are hydrolyzed by α– and β–amylases plus dextrinase to fermentable sugars (principally maltose), small oligosaccharides, and dextrins. The process of converting starch to fermentable sugar is called "mashing" in the brewing industry. Since malt is an expensive ingredient, a nonenzymatic source of starch or adjunct (for example, corn grits) is often added to the mash. Protein conversion to amino acids and peptides occurs only to a limited extent during mashing. The soluble extract (wort) is then filtered off from the spent grain and is pumped into the brew kettles, where it is boiled with hops to impart the characteristic bitter flavor of beer. Recently, an extract of hop inflorescences has been used as a substitute for boiling with hops. The spent grain residue is a useful by-product and is sold as an animal feed. After boiling the wort is cooled, usually by heat exchangers, to a low temperature suitable for inoculation with brewers yeast. Most beers are fermented by the bottom fermentation process (or lager fermentation), in which strains of the yeast *Saccharomyces carlsbergensis* are used. The term "bottom fermentation" refers to the tendency of this yeast to remain suspended for a limited time in the fermenting liquid, after which it settles or flocculates to the bottom. Fermentation is carried out at rather low temperatures, somewhere in the neighborhood of 10°C. The main fermentation, which lasts five to seven days, is followed by a lagering (or resting) period of several additional weeks at a temperature close to 0°C. During this time fermentation is completed, and yeast cells, as well as proteins insoluble at low temperature, settle out. The beer is then ready for final filtration, carbonation, and bottling.

The other principal type of fermentation, top fermentation, is conducted at higher temperatures, about 20–25°C, and is used mainly for the production of ales in Great Britain and Ireland. In this fermentation, strains of *Saccharomyces cerevisiae* are used, which tend to rise to the top of the fermenting liquid. Most of the yeast layer is skimmed off the surface when the fermentation is complete or subsides. Lager beers and ales contain about 4 and 6% alcohol, respectively.

Because of its multiplication during fermentation, brewer's yeast is a by-product in the brewing industry, and the yeast may be used in animal feed formulas because it is a valuable source of protein, amino acids, and vitamins. It is also used for therapeutic purposes in human nutrition, but the bitter flavor imparted by the hops that is concentrated in the yeast cells is a problem. A debittering process, which involves washing the yeast in a weakly alkaline solution, can remove part of the bitter flavor.

Sake Brewing

Sake or Japanese rice wine dates back to the ancient Orient. This beverage, which contains about 17% alcohol by volume, is made from rice. In sake brewing the rice starch is converted by amylolytic enzymes from the fungus *Aspergillus oryzae,* and the sugars thus formed are fermented by special strains of *Saccharomyces cerevisiae* that are usually designated as *Saccharomyces sake.* Briefly, the procedure involves the following steps. *Aspergillus oryzae* spores are inoculated in polished rice that has been steamed and cooled. The fungal development causes elaboration of amylases and proteases; this rice product with enzymatic activity is termed *koji.* A starter culture is then made in which steamed rice is supplied with koji, water, sake yeast, and some lactic acid for pH control and the prevention of growth of undesirable bacteria.

Fermentation ensues, and this starter culture is termed "seed mash" or *moto*. The moto culture is then added to the main mash or *moromi*, consisting of steamed rice, water, and additional koji. The main fermentation, which is done at 10–15°C for about one month, is followed by filtration, bottling, and pasteurization. In some operations lactic acid bacteria form part of the microflora in the main fermentation, where they produce lactic acid for pH control. It is customary to warm the sake to about 37°C before serving.

Wine

Wine is an alcoholic beverage made by fermentation of the juice of fruits or berries. The juices of these fruits already contain fermentable sugars, and yeast can be used directly for the fermentation of the sugar into alcohol. Wine yeasts are strains of the species *Saccharomyces cerevisiae*, which are selected by different wineries for their suitability in the production of specific varieties of wine. Recently, some strains of good wine yeasts have been available commercially in the form of active dry yeast. Such yeast can be rehydrated readily and can be a great asset to a winery, especially in the early production season before actively fermenting inocula from the winery itself are available. The use of natural fermentations, in which the yeasts present on fruits or berries are relied on for starting the fermentation, is not common today because there is a greater possibility of developing undesirable yeasts or bacteria in a less competitive environment.

Most wine is made from grapes. Grapes are harvested when they have reached the desired sugar and acid content and are then fed into a crusher, which crushes the berries and removes the stems. In the production of red wines the skins of dark grapes are allowed to remain in the fermentation tanks for several days, and the alcohol formed extracts the anthocyanin pigments from the skins. For white table wines, which are made from either white or dark grape varieties, the skins are separated at an early stage, usually after five to eight hours. Fermentation is continued until all the sugar is used up. The wine is then separated from yeast and other insoluble debris, which settle out (the sediment is termed "lees" in wineries), and allowed to age. Table wines contain approximately 12% alcohol by volume.

The true sherry wines of Spain (especially in the region near Jerez de la Frontera) are fermented in a different way. Special strains of yeast are used that rise to the surface of wine with about 15% alcohol in partially filled oak barrels and form a continuous deck (film or "flor") after the sugar is completely fermented. This subsequent oxidative stage is responsible for imparting the characteristic sherry bouquet to the wine over a period of many months or even years. During its action the yeast lowers the acid content but increases the level of acetaldehyde in the wine. Besides acetaldehyde, many aldehydes, esters, ketones, and higher alcohols characteristic of Spanish sherry have been identified by gas chromatographic techniques. These compounds are normally present in only trace quantities. After the fermentation is complete, the sherry is fortified with brandy to raise the alcohol content to approximately 17–19% by volume.

Some Special Beverage Fermentations

Several special fermentations are used that are often characteristic of certain regions or countries of the world.

In certain countries mead is a popular drink that is made by diluting honey and allowing it to ferment by a suitable wine yeast. A small amount of diammonium phosphate is sometimes added as a yeast food, since honey is rather low in yeast nutrients.

A number of unusual fermented beverages have been made for centuries in various countries; in some of these products bacteria and yeast act together to produce the required end product. One of these, the "Tibi" fermentation, has been studied in considerable detail. This popular Swiss drink is a sour, weakly alcoholic, carbonated liquid, made by the fermentation of a 15% cane sugar solution to which dried figs, raisins, and a little lemon juice have been added. The inoculation is done by adding a number of Tibi grains. These consist of a capsulated bacterium (*Betabacterium vermiforme*, presumably a capsule-forming lactobacillus) and a yeast (*Saccharomyces intermedius*, now considered a special strain of *Saccharomyces cerevisiae*) that live symbiotically. The combined action produces lactic acid, alcohol, and CO_2. The Tibi grains multiply during fermentation and can be transferred to a subsequent batch. The "ginger beer plant," which has been used in England for similar purposes and was described by Ward in 1892, probably is identical with the "Tibi Konsortium." The "tea fungus" is used in Indonesia to prepare an aromatic, slightly acid drink from tea infusion, to which 10% sugar has been added. Two symbionts make up the "fungus," *Acetobacter xylinum* (producing the acid) and *Saccharomycodes ludwigii*, a yeast that produces small amounts of CO_2 and alcohol. Other bacteria and yeasts appear to be associated with these last two organisms. Another fermented tea beverage is "Teekwass" (Russia), in which a symbiosis between an *Acetobacter* species and two yeast species has been reported. Several other fermented products are known where symbiosis between lactic acid bacteria and yeasts is involved (see sour-dough French bread under "Baker's Yeast"). The responsible organisms have been studied only superficially—largely at a time when satisfactory methods of identification were not available. A thorough reinvestigation of these fermentations would be worthwhile. Examples are Kefir fermentation (Caucasus), caused by inoculating Kefir grains into milk; Kumis (Asia); Leven (Egypt); and Mazun (Armenia).

Distilled Spirits

A spirit is a potable alcoholic beverage obtained by distilling an alcohol-containing liquid followed by treating the distillate to obtain a beverage of specific character. Examples include brandy, whiskeys, vodka, gin, rum, and tequila. Their alcohol content usually varies from 35–50% by volume.

Brandy is the distillate of wine or a fermented fruit juice. The distillate is aged for a number of years in oak casks. Brandies distilled from wine produced in the Cognac region of France are the only spirits entitled to the name "cognac". A special type of plum brandy produced in southeastern Europe is called "slivovitz."

Whiskeys are made by distilling fermented grain mashes followed by aging in oak casks. Scotch whisky is made from barley malt that has been treated with peat smoke, which is responsible in part for the characteristic flavor of this product. American bourbon whiskey is made from a corn or grain mash to which malt has been added for starch conversion (see Beer Brewing).

Vodka is made by distilling a fermented grain mash. The product is highly rectified during distillation, and thus is a very pure neutral spirit without a pronounced flavor. For greater palatability, some glycerol is added to the distillate. It is not aged.

Gin, like vodka, also consists of pure grain spirits to which has been added the distilled extract of juniper berries and certain other herbs. Gins are generally not aged.

Rum is the alcoholic distillate from fermented sugar cane juice or molasses. It contains various concentrations of congenerics (fusel alcohols, esters, and aldehydes). Dark rums are supplied with caramel to obtain the proper color. Rums are aged in oak barrels for various periods.

Tequila is a distilled beverage of Mexican origin

and is made from the fermented extract of certain species of agave (century plant). The product is aged for various periods.

PRODUCTION OF ETHANOL BY YEASTS

Industrial alcohol is made either by yeast fermentation of sugar solutions or by catalytic hydration of ethylene, a product from the petroleum industry. The world's potable alcohol is made by fermentation because of government regulations. Over the years the industrial production of ethanol by yeast fermentation has fluctuated widely, primarily in a declining direction because of competition from ethylene-produced synthetic ethanol. Recently, considerable scarcities of ethanol have been developing worldwide. This has been caused by serious crop failures, thus limiting the amount of starch or sugar available for this process, and by shortages and soaring prices of ethylene feed stock. Of the roughly 300 million gallons of industrial alcohol produced in the United States in 1974, only 10 million gallons were made by fermentation. However with ethylene prices rising and the diversion of raw material into more profitable products, such as synthetic fibers and plastics, the fermentation alcohol industry is likely to grow again, provided sufficient sources of polysaccharides or sugar are available for this purpose.

The fermentation is carried out by selected strains of *Saccharomyces cerevisiae* (distiller's yeasts) because of their rapid growth rate in the presence of moderately high concentrations of alcohol. The theoretical conversion of glucose to ethanol and CO_2 is 51% and 49%, respectively, on a weight basis. In practice the conversion is 90–95% of the theoretical values, depending to a large extent on medium composition—especially the source of nitrogen. At the completion of fermentation, the ethanol concentration is approximately 8%; it is then concentrated and purified in several distillation and rectifying columns to 95% ethanol, the maximal concentration attainable by distillation.

Two principal substrates are used. The first one is molasses, a by-product of the sugar industry, which contains primarily sucrose and some invert sugar. Fermentation of this substrate is relatively simple, since about 90% of the sugars in molasses are fermentable by yeasts. The highly concentrated molasses is diluted with water to a sugar content of 14–18% and pumped into the fermenter. An actively fermenting starter culture is also added. Some nitrogen (in the form of ammonium ion) and other minerals are usually added to supplement those already present in molasses. Both batch and continuous fermentations are used. In the batch process, fermentation is usually complete after thirty-six to seventy-two hours. It takes 2.3–2.7 gallons of black-strap molasses (cane sugar molasses) to produce 1 gallon of 190 proof (95%) alcohol.

The second major raw material is cereal grain. Since distiller's yeast cannot ferment starch, the milled grain slurry is cooked to hydrate and gelatinize the starch, cooled, and supplied with a special type of distiller's barley malt that has a high diastatic (amylolytic) potency. The amylase enzymes of the malt hydrolze the starch at a temperature of 62–64°C to fermentable sugars and low molecular weight dextrins. After sufficient conversion, the mash is cooled in heat exchangers to a temperature ranging between 21 and 32°C and inoculated with yeast. During fermentation additional conversion of the dextrins by residual malt enzymes occurs. To supply the yeast with nutrients in addition to those supplied by the grain, about 20–25% stillage (the residue of the distillation columns) is added, which is very rich in vitamins, minerals, and nitrogenous substances. Stillage also reduces the pH of the mash to a more favorable level (about 4.8–5.0). Fermentation is complete after approximately three days. The liquid, containing insoluble grain residue and yeast, is then taken to the stills for concentration and purification of the alcohol.

A number of other raw materials have been used occasionally. For example, wood wastes, such as sawdust, can be hydrolyzed by heating in the presence of strong acid, and the resulting hexose sugars can be fermented after neutralization of the acid. However, since wood is rich in pentose sugars, which cannot be fermented by yeasts, this raw material is not as suitable as the carbohydrates from grain.

Hydrolysis of cellulose to glucose by cellulolytic enzymes from fungi is currently under study in some laboratories. If such a conversion can be made efficient, a large quantity of glucose could be generated from waste cellulose products such as old newsprint.

A recently announced new development in industrial alcohol production via fermentation involves fermentation of media with much higher sugar concentrations than the customary 12–14%. Since this procedure would lead to higher alcohol levels and consequently slow down the fermentation rate toward the end, the fermentation process occurs in a vacuum. Through this technique, the alcohol is continually removed by distillation, and thus does not reach concentrations inhibitory to the yeast. The vacuum evaporation simultaneously causes an automatic cooling of the mash. This saves in cooling cost of a traditional fermentation vessel where heat generated by the alcoholic fermentation raises the temperature of the medium. Heavy inocula of yeast are used in this process to speed up the fermentation.

GLYCEROL AND OTHER POLYHYDROXY ALCOHOLS

The formation of glycerol by wine yeasts and other species of *Saccharomyces* has been known for many years. Normally about 2–3% of the weight of the sugar fermented consists of this trihydric alcohol. Because of the great demand in Germany for glycerol during World War I, Neuberg developed two processes that greatly increased the yield of glycerol by strains of *Saccharomyces cerevisiae*. The normal alcoholic fermentation is often called Neuberg's "first form" of fermentation. His "second" and "third forms" are those in which steering agents (bisulfite and alkali, respectively) are used to divert the fermentation into different channels.

The function of bisulfite is its ability to combine with acetaldehyde, one of the last intermediary compounds in the pathway of alcoholic fermentation, and thus to interfere with its reduction to ethyl alcohol. The reduction then occurs with dihydroxyacetone phosphate as the substrate, an earlier intermediate in the scheme of alcoholic fermentation. The result is formation of glycerol phosphate, which is followed by enzymatic hydrolysis of the phosphate group in the yeast cell and formation of free glycerol. Because of the toxicity of sulfite, the growing conditions are quite critical. A reducing sugar level of 20–22% and a free sulfite content of 3–3.5% in a neutral medium has been found best. Such a medium can yield 27% glycerol on the basis of the sugar supplied. The difficulty is the recovery of glycerol (by vacuum distillation) from the fermented broth, which contains quite large amounts of salt.

Neuberg's third form of fermentation, or the alkaline fermentation, is based on the addition of high concentrations of sodium carbonate, giving an alkaline reaction. Under such conditions acetaldehyde undergoes a dismutation reaction, which is an oxidation reduction resulting in 1 mole of acetate and 1 mole of alcohol from 2 moles of acetaldehyde. Again, acetaldehyde is prevented from being reduced by the hydrogen from reduced coenzyme (NADH), which instead reduces dihydroxyacetone phosphate. With suitable strains of *Saccharomyces cerevisiae* (those that are adapted to alkaline conditions), 10–24% of the fermentable hexose sugar can be converted to glycerol. As with sulfite, the recovery is very difficult due to the presence of large amounts of sodium carbonate.

Of considerable interest is the discovery that under highly aerobic conditions many haploid, osmoduric (sugar- or salt-tolerant) species of *Saccharomyces*, *Pichia*, and related imperfect yeasts belonging to *Torulopsis* produce, in the absence of steering agents, various combinations of glycerol, erythritol, D–arabitol, and sometimes D–mannitol. Under optimal conditions, as much as 40% of the glucose can be converted to polyhydric alcohols. A low phosphate level in the medium favors the synthesis of these compounds, which are thought to be formed by dephosphorylation of phosphorylated intermediates of the pentose cycle, followed by reduction by specific NADPH-dependent dehydrogenases. Reduced coenzyme is regenerated during the early steps of glucose–6–phosphate oxidation to ribulose–5–phosphate.

Yeast species can be grouped by the products they synthesize. These categories include the following: only glycerol; only erythritol; only D–arabitol; erythritol plus D–arabitol; glycerol plus D–arabitol; or glycerol, D–arabitol, and erythritol. Thus by selecting the proper yeast one can produce almost any combination of these substances. Recovering these products by distillation or crystallization is much easier than from Neuberg's alkaline or bisulfite fermentation.

When once again microbial production of glycerol is considered economical, or if a greater market should develop for other polyhydric alcohols, use will undoubtedly be made of selected haploid *Saccharomyces*, *Pichia*, or *Torulopsis* species, which lend themselves extremely well to this purpose.

BAKER'S YEAST

In the modern baker's yeast industry, specially selected strains of *Saccharomyces cerevisiae* are used. The product is available on the market in two forms: as a compressed yeast cake of approximately 70% moisture content, and as active dry baker's yeast of approximately 7.5% moisture content.

Yeast has been used for many centuries for the leavening of bread. Prior to Pasteur's time, when the role of yeast in bread leavening was not clearly understood, the yeast was propagated from dough to dough or from mash to mash. Later, surplus distiller's yeast (after pressing out excess water) was commonly used. Brewer's yeast usually had inferior leavening powers and left a bitter taste from the hops used in brewing.

The earliest production of baker's yeast as a primary product was by growing it in a grain mash under conditions that were only moderately aerobic. As a result, yields of yeast were low, and much alcohol was produced as a by-product. Four major fundamental discoveries are the basis of the modern baker's yeast industry.

The first is the use of molasses instead of grains as the carbon source. This by-product of the sugar industry contains a very high concentration of utilizable sugar and, in addition, has certain growth factors and minerals. Second, yeast can be grown efficiently with an inorganic source of nitrogen, added as a mixture of ammonium sulfate and ammonium hydroxide—the latter for the purpose of balancing pH changes. Third, based on the discovery of Pasteur that aeration inhibits alcohol formation, fermentations today are carried out with extremely efficient aeration so that alcohol production is almost completely suppressed. Fourth, the use of feeding schedules allows the rate of substrate addition to follow the growth curve of the yeast. A low sugar concentration is maintained throughout the process, helping to induce respiratory activity, to suppress alcohol formation, and to increase the yield of yeast greatly (see Chapter 6, the Crabtree effect).

Baker's yeast is produced from pure cultures through many stages of increasing volume. Before the last stage is initiated in the largest fermentation tank, the yeast from the previous stage is concentrated by centrifugation to allow the use of a large inoculum. After a sufficient degree of multiplication (usually three or four generations) has been reached, the supply of nitrogen is stopped, but aeration and carbohydrate addition are continued to "ripen" the yeast. During this period young buds mature, and the yeast increases its content of reserve carbohydrates (mainly glycogen and the disaccharide trehalose). This ripening process results in a greater storage stability than if immature buds were present in the yeast. Finally, the product is harvested, washed in centrifuges, and pressed dry in filter presses; it is then extruded from a machine and packed in blocks of specified weight. These final procedures are all carried out at low temperatures, and the yeast is stored at approximately 0°C, since it is quite susceptible to breakdown (autolysis) and to microbial spoilage.

Because of the susceptibility of compressed yeast to spoilage, active dry yeast for baking has been developed over a period of years. Different strains of *Saccharomyces cerevisiae* are used, and the yeast is treated in its final growth stages so that the reserve carbohydrate content becomes very high. The harvested yeast is dried under very careful conditions of temperature and humidity, so that the sensitive enzymes responsible for fermentation are not damaged during the drying process. Smooth, round pellets of dried yeast are made from crumbled, compressed yeast in revolving steel drums through which warm air flows. Dried baker's yeast is also marketed in short strands of porous appearance. These are produced by drying on a slowly moving steel-mesh belt in an air stream. The yeast is dried to a critical moisture content of approximately 7.5%, because only at this moisture level can the yeast be rehydrated to nearly the original fermentative activity in the home or in the baking industry. In addition, rehydration of the dried yeast must be done in water of about 43°C. Lower temperatures cause extensive leaching of the cell contents, and a loss in baking quality results. Active dry baker's yeast is of such quality today that its baking properties are only a little lower than those of compressed yeast. Moreover, the long-term storage stability of active dry yeast is much superior to that of compressed yeast, provided it is stored in vacuum or in a gas atmosphere free of oxygen.

The French Sour Dough Process

The microbiology and technology of this process are quite different from procedures used for ordinary bread baking. Bakeries making sour dough bread generally inoculate the dough with a starter, or mother sponge, that is continually propagated in the bakery, which is the source of both the leavening and souring powers. It contains a mixture of a yeast (*Candida milleri*) and a heterofermentative lactic acid bacterium (*Lactobacillus sanfrancisco*) that propagate side by side in the dough. The reason for their symbiotic growth is that the yeast utilizes glucose but (in contrast to *S. cerevisiae*) not the maltose in the dough, whereas the bacterium does the reverse; thus, there is no competition for a carbon source in the medium. The lactobacillus produces lactic and acetic acids plus some CO_2 from maltose; *Candida milleri*, which is unusually tolerant to acetic acid, carries out alcoholic fermentation at the expense of glucose; this is responsible for the leavening of the dough.

YEASTS IN FEEDING

Because vast areas of the world do not produce sufficient protein to feed its population effectively, much work has been devoted to attempts to supplement the available proteins with those of microorganisms. Because microorganisms can very rapidly convert inorganic nitrogen into amino acids and proteins, this approach holds a considerable advantage over protein from higher animals or plants. In addition, microorganisms—especially yeasts—form a rich source of many of the B-group vitamins. Although yeast contains all of the essential amino acids needed by higher animals and man, the sulfur-containing amino acids L–cysteine and L–methionine, as well as tryptophan occur in too low a concentration. Consequently, microbial proteins or single cell proteins (SCP) are most effective as supplements (up to 50%) in animal diets rather than as the sole source of protein. Also, because of digestion problems with whole yeast cells, not all of the potentially available amino acids are utilizable by animals.

The most widely used yeast is *Candida utilis (Torulopsis utilis*, or torula yeast), the proteins of which are quite valuable. For example, 70% of the lysine content in its protein is available to chickens, as compared with 100% for the pure amino acid. All evidence points to highly beneficial results when a limited fraction of the dietary protein is supplied by yeast. For human nutrition there is a problem of susceptibility to a high percentage of yeast protein in the diet, to which at least some individuals develop gastrointestinal disturbances. Other objections include the tough cell wall of yeasts, the high nucleic acid content, and the lack of texture or a slimy texture. In connection with the high purine content of yeast cells (yeast may contain up to 12 or even 15% RNA), it should be noted that mammals other than man contain uricase. This enzyme is able to convert uric acid (a product of purine metabolism) into the soluble and easily excretable metabolite allantoin. Since man lacks this enzyme, purines are not converted beyond uric acid. Thus, there is a danger of uric acid accumulation in the tissues and joints analagous to the conditions in gout. Also, stones may be formed in the kidneys and bladder. This has led to considerable research of reducing the RNA content of yeast. One of the more promising processes is heat activation of the endogenous RNAase of yeast that breaks down the RNA to nucleotides, which can be extracted from the yeast without losing appreciable amounts of protein. Heating to 68°C for several seconds seems to disrupt the ribosomes. If the yeast is subsequently incubated for several hours at 45–55°C, the nucleic acid content can be reduced to approximately 1–1.5%. Other research currently in progress involves the disruption of yeast cells by mechanical means, extracting the proteins, and processing them into a more textured product with a higher acceptability than the relatively bland, textureless product made directly from the yeast cell.

Some of the advantages of SCP over proteins from plants and animals are the very short generation time of the microbial cells; their high protein content, usually 40–50% on the dry weight basis; the use of cheap substrates that are available in large quantities; small land areas; and that microorganisms can be easily and rapidly modified genetically for superior production. A disadvantage is that SCP processes are extremely capital intensive.

One of the important conditions for successful SCP

production is the proper combination of a cheap, abundantly available substrate for growth and a microbial species that will grow on it.

In tropical areas of the world, starches are abundantly available—for example, in the form of tapioca and cassava roots. It is usually thought that yeasts are unable to grow on this substrate without its previous hydrolysis, and this is true for most species. However, there are exceptions; we have been involved in pilot-scale experiments in which species of the genus *Schwanniomyces* were grown on cassava root starch, which they can effectively utilize after the starch has been gelatinized and mineral nutrients added. In this way an enriched food, originally almost completely devoid of protein, can be made in a relatively simple fermentation.

More commonly, yeast itself is grown and harvested from abundantly available substrates, such as sulfite waste liquor, a by-product from the paper industry, or on crude or purified petroleum products. The use of sulfite waste liquor as a substrate with a suitable yeast (such as *Candida utilis*) that can utilize the sugars (including pentoses) formed from the hydrolysis of hemicelluloses of wood pulp has the advantage of reducing environmental pollution besides being a very cheap substrate. Moreover, this species requires no added vitamins in the medium of growth. The yeast is harvested, washed, and dried at a relatively high temperature and is then incorporated in animal feed rations.

Much current research is concerned with the enzymatic conversion of cellulosic substrates (for example, agricultural by-products such as bagasse and wood products) to fermentable sugars. To make these substrates more susceptible to the cellulase complex of enzymes, they usually have to be pretreated with alkali or acid. The enzymes are normally obtained from fungi (for example, *Trichoderma viride* and *Aspergillus* species) grown on cellulosic substrates.

Following the discovery that many species of yeasts —particularly species of *Candida* [for examples, *C. tropicalis*, *C. maltosa*, and *C. lipolytica* (= *Saccharomycopsis lipolytica*)] —can utilize the normal or straight chain alkane fractions of petroleum products, much research in this area was initiated. Since gas oil (a liquid petroleum fraction boiling between approximately 225 and 380°C) contains only about 10–25% straight chain paraffinic hydrocarbons, and since this substrate may be contaminated with possible aromatic carcinogens, it is generally considered preferable to manufacture normal paraffins by the molecular sieve process and use this highly purified, odorless substrate rather than the crude oil fractions. However, this process is quite costly and may be impracticable on a commercial scale.

More recently, attention has been given to methanol and ethanol as carbon sources for yeast. These compounds are produced by catalytic oxidation of the corresponding hydrocarbons methane and ethane. Although the literature primarily mentions bacteria as methanol utilizers, there are a number of species of yeast (particularly in the genera *Hansenula, Pichia, Candida,* and *Torulopsis)* that grow vigorously on methanol as the sole carbon source. An advantage of these substrates is that their purification is not expensive: they are available in highly purified form without the danger of contamination by carcinogens found in the original oil product. A further advantage of alcohols as carbon sources over hydrocarbons is that the substrate is partially oxidized. Therefore, less heat is evolved in further oxidation during growth, and less cooling of the fermentation tanks is needed. Finally, these substrates are miscible with water—another great advantage over normal paraffins. A problem may be their availability and extreme sensitivity of SCP processes to the cost of these raw materials.

LIPID MATERIALS

Some species of yeast produce remarkably high levels of lipid materials, which are quite evident when such yeasts are observed under the microscope. The cells contain one or more rather transparent, spherical fat globules of uneven size. They tend to cause light refraction as if they were small lenses, appearing bright in strong light.

Commercial production of lipid materials by yeasts has not been economical, except during periods of war. *Trichosporon pullulans* (synonym *Endomyces vernalis),* species of *Lipomyces* (one species formerly called *Torulopsis lipofera), Metschnikowia reukaufii, Oospora lactis,* and *Rhodotorula* species are examples of yeasts that have been used experimentally or commercially for fat production. These organisms have primarily a respiratory (oxidative) metabolism, and they can convert carbohydrates to fat. The nutrient conditions must be such that protein synthesis is limited, as when the nitrogen or phosphate content of the medium is low. Under such conditions the crop of cells is not large, but the fat content is very high (up to 50–60% of the dry weight of the cells). Commercial production is done with compromise nitrogen levels that result in a somewhat lower fat content but a higher cell crop. In this approach there is a growth phase in a rich medium, followed by a "fattening" phase in a nitrogen-deficient medium. More recently, a yeast has been reported (*Cryptococcus terricolus*, but now considered a strain of *Cryptococcus albidus*) that produces high concentrations of fat regardless of the nitrogen content of the medium. The lipids include triglycerides, fatty acids, phospholipids, and sterols. The bulk of the fatty acids are unsaturated, which is responsible for the fats having a low melting point and being liquids at room temperature. Ergosterol is the main sterol of yeast, comparable with cholesterol produced by animals. The fat coefficient, or the number of grams of fat produced from 100 grams of sugar, can be as high as 15 to 18 under suitable conditions.

VITAMIN PRODUCTION BY YEASTS

Many yeasts, among them most species of *Saccharomyces*, require one or more vitamins in media to supplement the nutrients that a yeast can synthesize itself. Other yeasts (for example, *Candida utilis* and *Hansenula anomala)* can synthesize all the necessary vitamins from simple precursors. *Saccharomyces cerevisiae* is able to concentrate and absorb large amounts of thiamine, nicotinic acid, and biotin from the medium of growth, and thus form enriched products. By supplementing the medium with crude sources of these particular vitamins (or sometimes even their precursors), one can make yeast of exceptionally high vitamin content. Such products are sold for therapeutic purposes.

Baker's yeast, brewer's yeast, and *Candida utilis* are good sources of thiamine, riboflavin, pantothenic acid, nicotinic acid, pyridoxine, folic acid, biotin, *para*-aminobenzoic acid, inositol, and choline.

Some strains of the yeast-like organisms *Eremothecium ashbyi* and *Ashbya gossypii* can oversynthesize extremely large amounts of riboflavin. This property has been used to produce inexpensive, crude sources of riboflavin for use in animal feed formulas. The last species is genetically the most stable in its ability to overproduce this vitamin. Both organisms grow optimally under aerobic conditions at 26–28°C. Common sugars or, even better, lipids (for example, corn oil or soybean oil) serve as carbon sources. A crude organic nitrogen source is required for high yields (for example, an enzymatic digest of collagenous proteins). Accessory factors are supplied by adding corn steep liquor, yeast extract, or distillers solubles. Glycine and a nonionic surface active agent (for example, Tween 80) further increase the yield. Most of the riboflavin is synthesized after active growth ceases. Yields up to 5 g/l have been reported in six to seven days of growth. With such high yields, microbial riboflavin can compete well with the synthetically produced vitamin.

Some of the sterols produced by yeast can be transformed, upon irradiation with ultraviolet light, into vitamin D_2 (calciferol). Selected strains of *Saccharomyces* can produce 7–10% ergosterol on the dry weight basis, and such strains can be irradiated with ultraviolet light to produce a yeast of superior vitamin D content.

Although some strains of yeast belonging to the genera *Cryptococcus* and *Rhodotorula* can synthesize significant quantities of β–carotene, the concentration is too low to be economically important. Production of β–carotene by fungi belonging to *Zygomycetes* (for example, *Blakeslea trispora*) is more promising than by yeasts.

YEASTS AS SOURCES OF ENZYMES

Baker's yeast (*Saccharomyces cerevisiae*) is very rich in invertase, an enzyme that splits sucrose into a mixture of glucose and fructose (invert sugar). When grown aerobically on sucrose, the enzyme is induced. This yeast may contain 50 to 100 times the concentration of invertase needed to maintain fermentation at its maximal rate. Invertase, a mannan-containing glycoprotein, is associated with the cell wall and can be obtained in soluble form by treating the cells with toluene or chloroform, followed by autolysis. Purified invertase is used in the production of artificial honey, invert sugar, and in the manufacture of cream-center bonbons.

The enzyme lactase is primarily made from *Kluyveromyces fragilis*, a lactose-fermenting yeast. The inducible enzyme obtained from cell extracts is used to reduce the lactose content of dairy products. Reduction of the lactose content in ice cream prevents crystallization of the sugar. In other dairy products it removes the lactose to which a very significant proportion of the world's population is sensitive because of the absence of lactase in the intestinal tract. In such individuals the lactose is subject to microbial fermentation in the lower intestine, and as a result gastrointestinal disturbances occur.

POLYSACCHARIDES

Recently, considerable interest has developed in the use of microbial polysaccharides for various industrial purposes, including their use in crude oil recovery and as modifiers of texture and consistency in food products. Phosphomannans, which are produced by certain species of the genus *Hansenula*, *Pichia*, and *Pachysolen* are among the most interesting and promising polysaccharides. These yeasts have the remarkable ability to convert up to 50% of the glucose in the medium to extracellular capsular polysaccharides, which are subsequently released into the medium. The polymers contain only mannose in addition to phosphate. Depending on the species of yeast producing this polysaccharide and medium composition, the ratio between mannose and phosphate can vary from 2—3, up to 10—27 mannose units per phosphate unit. Isolated phosphomannans form highly viscous, clear solutions that are quite resistant to bacterial attack. They have been used experimentally for controlling the consistency of various foods.

11 / Yeast Classification

In order to report biological or biochemical properties of a yeast (or of any other living organism), a prime requisite for the microbiologist is the ability to recognize, identify, and name the organism with which he or she is concerned. To facilitate recognition of organisms by persons with different native tongues, every plant or animal or microbial taxon (taxonomic group) has an internationally recognized scientific name taken from the Latin (or latinized Greek). In contrast to many animals and plants, very few yeasts have vernacular names, such as baker's yeast (*Saccharomyces cerevisiae*).

At present three independent international codes of nomenclature exist, namely for plants, animals, and bacteria. Yeasts (as do other fungi) fall under the botanical rather than the bacteriological code. The botanical code prescribes that no fungus (or yeast) name can be recognized botanically, if published since January 1, 1935, unless a description in Latin is given. Names given without such a description are not validly published and have no standing (a nomen nudum). A second requirement is that the scientific name of a species consist of two parts—a binomial combination. The first component represents the name of the genus in which the species is included, while the second component is the specific epithet or species name. Together they represent the complete name of a species. In the professional literature, names of yeasts should be followed (at least once in a publication) by the name or names of the investigator(s) responsible for the original description of the organism, and this in turn should be followed by the date of publication—for example, *Schizosaccharomyces octosporus* Beijernick 1894. The purpose of this practice is to identify unequivocally the species and its authority. The botanical code contains many additional rules and recommendations, which are modified periodically during the various international botanical or mycological congresses. For further details consult the latest edition of the *International Code of Botanical Nomenclature.*

Occasionally, a species is transferred to a genus different from the one it was first placed in when new knowledge justifies such a decision. This could happen, for example, when the original observations on spore morphology prove to be incorrect and the yeast therefore belongs in a different genus (for example, *Debaryomyces vini* was transferred to the genus *Pi-*

chia as *P. vini* based on reevaluation of its spore surface topography). If this is done, the species name remains the same, unless the genus already contains a species with the same specific epithet. Obviously, no genus could have more than one species with the same name.

The following are two other examples: *Pichia membranaefaciens* Hansen 1888 was originally described as *Saccharomyces membranaefaciens* by Hansen, but was later transferred to the new genus *Pichia* by Hansen himself when it became desirable to exclude yeasts from the genus *Saccharomyces* that did not ferment significantly and formed surface pellicles in liquid media. *Saccharomyces apiculatus* Reess 1870 was changed to *Kloeckera apiculata (Reess emend.* Kloecker) Janke 1870. The term *emend.* signifies that Kloecker gave a more comprehensive definition of the species than Reess was able to do in 1870, and thus he amended the description. Janke, on the other hand, transferred the organism from the genus *Saccharomyces* to *Kloeckera* because ascospores were not formed, and the cells were largely apiculate (lemon shaped) rather than ellipsoidal. This particular transfer required a change in the ending of the specific epithet because *Saccharomyces* is a masculine noun and *Kloeckera* is feminine. Although Janke coined the name *Kloeckera* in 1928, the original date of discovery (1870) remains attached to the name of the species.

The basis on which species are separated from related ones is not always uniform for sporogenous and asporogenous yeasts. Generally , for both groups most attention is paid to physiological differences, such as the ability to ferment various sugars, to assimilate certain carbon compounds, to grow on ethylamine and nitrate as sole nitrogen sources (the latter especially for asporogenous yeasts), the maximum temperature for growth, the resistance to cycloheximide, and the osmotolerance.

Many controversies exist in viewpoints concerning biological taxonomy. Some investigators are "splitters," those inclined to establish many species on the basis of relatively minor differences, and others are "lumpers," wishing to reduce the number of species.

As in all other groups of living organisms, attempts are being made to classify yeasts into families, genera, and species on the basis of natural relationships, thus taking into account that yeasts (like higher plants and animals) have developed into their present state by evolutionary processes. Hence, there are primitive species, in an evolutionary sense, that developed long ago, and more advanced species that developed more recently. A study of the development of evolutionary lines (phylogeny) in higher plants and animals is dependent, in part, upon the discovery and study of fossil remains. The simple structures of most microorganisms, unfortunately, do not lend themselves very well to a study of their fossil predecessors, even though these forms have been found in deposits of the oldest geologic eras. Instead, evolutionary lines and natural relationships in yeast have been deduced from their associations with various primitive or advanced higher plants or animals (thus their ecology, see Chapter 8), from structural differences in the mannan component of the cell wall, from differences in the structure of coenzyme Q (ubiquinone), and from comparisons of nuclear genomes including nucleic acid hybridizations.

Several examples of such phylogenetic deductions are given here, but an in-depth consideration of this complex field is beyond the scope of our book.

Based on his extensive studies of the genus *Hansenula,* Wickerham postulated that species associated with bark beetles parasitizing coniferous trees are the most primitive members of that genus. Subsequently, species evolved dependent on broad-leaved trees, and finally as free-living species, such as those occurring in soil or as contaminants of food or in industrial fermentations.

Differences in the chemical structure of the mannan component of the yeast cell wall have been considered by others as a useful indicator of common phylogeny and for the differentiation of species and higher taxa. Because of the complexities of direct chemical analyses as conducted by Ballou and coworkers, comparisons are usually done by indirect means. These include serological characterization of the mannan component as proposed by Tsuchiya and others, and its analysis by the technique of proton magnetic resonance (PMR) spectroscopy as advocated by Gorin and Spencer. Although serological characterization of yeast mannan has been helpful in determining relationships among various yeasts, a drawback is that many immunochemical determinants of yeast mannan are controlled by single genes. Gorin and Spencer established a number of PMR spectral types of yeast mannan. Although several PMR groups were found to correspond well with groupings based on currently used taxonomic criteria, often seemingly unrelated organisms were assigned to the same PMR-determined group.

Analysis of the coenzyme Q system (ubiquinone) by Yamada and Kondo has revealed that there are several distinct types in various yeasts. The different coenzyme Q types, which function in the electron transport system, vary in the number of isoprenoid units in the side chain of the molecule. They have been placed in several clearly defined classes that correspond closely to several accepted genera. Determination of the coenzyme Q type of some yeasts of doubtful taxonomic status has aided in their placement in appropriate genera. A positive correlation exists between the ability of yeasts to degrade n–alkanes and the type of coenzyme Q they possess (Q_9 in this case).

Molecular taxonomy based on comparisons of nuclear genomes in different yeast species has contributed greatly to a critical evaluation of currently used taxonomic criteria as well as to an understanding of the relationship between some species of yeast. In spite of approximately fifty morphological and physiological criteria now used for the characterization and differentiation of yeast species, it is not uncommon to find wide intraspecific variation in mole percent guanine plus cytosine in the nuclear DNA (mol % GC) of a number of presently accepted yeast species—particularly among the imperfect yeasts where the characteristics of a sexual cycle cannot be used for systematic purposes. Strains of such "species" therefore represent mixtures of species that cannot be distinguished by currently used criteria. Conversely, the indiscriminate use of a large number of (and often trivial) criteria has led to the description of many new "species," sometimes differing by only a single biochemical ability—for example, hydrolysis of a disaccharide. In many of the latter cases, "species" separated on this basis not only were found to have the same nuclear DNA base composition (mol% GC) but also a very high degree of DNA–DNA complementarity (homology) or DNA–DNA renaturation ability. Such "species" often differ by only a single gene, or possibly even by a single base pair in the case of point mutations. Thus, such yeasts with nearly identical nuclear genomes do not represent separate species, but, at best, they should be regarded as varieties or biochemical strains of a parental species.

Other approaches to the determination of relationships between yeasts are based on studies of the complementarity of the nuclear DNA and ribosomal RNA. Although only a small part (about 2%) of the nuclear genome codes for ribosomal RNA, there is evidence that this portion of the genome is more strongly conserved than the remainder. This approach, therefore, makes it possible to compare more distantly related species than is possible by determining whole DNA/DNA complementarity. Other investigators are studying the comparative structure of certain universal enzymes (proteins) in yeast species to determine the relative extent of divergence due to evolutionary forces.

The base composition of the nuclear DNA has been determined for most yeast species, and the range extends from about 27–70 mol % GC. The basidiomycetous species range from about 48–70% GC, and the ascomycetous yeasts occupy the lower range. Some small overlap of species occurs in the transition zone between about 47–50% GC. This knowledge has been useful in determining the presumed affinity of asporogenous species to the asco- and basidiomycetous classes of yeast.

In the differentiation of larger taxa (genera, families, orders, and classes) details of sexual and asexual reproduction, as well as cell and spore morphology play a much greater role than do physiologic properties (see Chapters 2 and 4).

The brief introduction to yeast systematics presented here illustrates the problems facing the taxonomist. In spite of many weaknesses, the currently available taxonomic keys give the reasonably trained yeast worker much better tools in the identification of cultures isolated from various sources than was possible thirty years ago. New criteria for differentiation and resulting improvements in these keys will make the task even easier. Our book is limited to a listing of the principal features of the yeast genera recognized by most workers in the field. This information is in Appendix A; those who encounter names of organisms in the text can orient themselves with respect to the features characterizing the genera in which these species have been placed. The arrangement of yeast genera differs in some details from other schemes, since it represents our views based on current knowledge.

In asco- and basidiomycetous yeasts the presence of a sexual life cycle permits a more reasonable grouping along "natural" lines than was possible in the past without this information. We therefore present an arrangement of the genera into subfamilies where information permits it. The asporogenous yeasts are placed, for convenience, in "form genera." Here it is recognized that some species may be placed in a single form genus on the basis of similarities in cell morphology and of certain biochemical properties. If their sexual cycles were known, however, such species might be found widely separated in their relationships. The ballistosporogenous genera are included among the asporogenous genera, even though some species are known to belong to the basidiomycetous yeasts.

Finally, in Appendix B all of the genera of yeasts and closely related organisms are listed alphabetically in very brief form. This tabulation is a glossary of the major features of the various genera. We feel that the examples cited in the book will be far more meaningful if the reader can rapidly orient himself by this guide. In Appendix A, however, the genera are listed by relationship rather than alphabetically.

Appendixes
Selected Bibliography
Glossary
Index

Appendix A / A Listing of Yeast Genera and Diagnoses.

An asterisk indicates the genera usually considered to be "yeast-like."

KINGDOM—FUNGI

  DIVISION—EUMYCOTA

    SUBDIVISION—ASCOMYCOTINA

      Class—Hemiascomycetes

        Order—Endomycetales

          Family—Endomycetaceae

              Genus—*Eremascus**

                  *Endomyces**

          Family—Saccharomycetaceae

            Subfamily—A. Schizosaccharomycetoideae

                Genus—*Schizosaccharomyces*

              B. Saccharomycetoideae

                Genus—*Dekkera*

                  Saccharomycopsis

                  (= *Endomycopsis*)

Arthroascus
Saccharomyces
Kluyveromyces
Schwanniomyces
Debaryomyces
Citeromyces
Pichia
Hansenula
Pachysolen
Cyniclomyces
(= *Saccharomycopsis*)
Lodderomyces
Wingea
Wickerhamiella

C. Nadsonioideae
Genus—*Nadsonia*
Hanseniaspora
Saccharomycodes
Wickerhamia

D. Lipomycetoideae
Genus—*Lipomyces*

Family—Spermophthoraceae
Genus—*Metschnikowia*
Nematospora
Coccidiascus
*Ashbya**
*Eremothecium**
(= *Crebrothecium*)

SUBDIVISION—BASIDIOMYCOTINA
Class—Teliomycetes
Genus—*Leucosporidium*

Rhodosporidium
Filobasidium
Filobasidiella
Aessosporon
Sporidiobolus

SUBDIVISION—DEUTEROMYCOTINA

Family—Sporobolomycetaceae

Genus—*Sporobolomyces*
Bullera

Family—Cryptococcaceae

Genus—*Cryptococcus*
Rhodotorula
Phaffia
Pityrosporum
Schizoblastosporion
Kloeckera
Trigonopsis
Brettanomyces
Torulopsis
Candida
Trichosporon
Oosporidium
Sterigmatomyces
Sympodiomyces

YEAST-LIKE ORGANISMS (other than those indicated above by *)

Aureobasidium
(= *Pullularia*)
Geotrichum
Taphrina
Prototheca
(a colorless alga)

The group of yeast-like organisms are "related" to the yeast genera in some cases by their frequent association with yeasts in their environment, and/or by their superficial resemblance to yeasts in various morphological and physiological properties. Those genera included in the taxonomic scheme indicated with an asterisk are normally not included within the "scope" of yeast genera, but rather are included as "yeast-like." In part, these views reflect phylogenetic schemes postulated by several early taxonomists.

Although detailed descriptions of each of the genera are given in various other publications, particularly in the monograph edited by Lodder (1970) and in more recent taxonomic papers, descriptions are given here that include the salient features distinguishing one from the other. In general, the description includes general characteristics of the vegetative structure (thallus), means of asexual reproduction, characteristics of the sexual structure (if present), general physiological properties, and any special features characteristic of the genus.

ASCOSPOROGENOUS YEASTS

Eremascus

Species of this yeast-related genus grow only as mycelium with cross walls; the mycelium does not disarticulate into arthrospores. Young cells may be multinucleate; older cells are uninucleate. Thick-walled chlamydospores may be either terminal or intercalary. Sexual reproduction is between gametangia of similar sizes, which may originate from adjacent cells of the same hypha or from cells of neighboring hyphae. The gametangial tips may coil around each other. After fusion of the gametangial tips, a spheroid- to ovoid-shaped ascus arises, containing eight spheroidal to ellipsoidal ascospores. Occasionally, asci may develop without apparent conjugation. Fermentative ability is lacking; growth in liquid media is in the form of a surface pellicle. Nitrate is not assimilated. Species of this genus are osmophilic and have been isolated from food products with a low water activity.

Endomyces

The vegatative body is a septate mycelium, reproducing by fission and by disarticulation into arthrospores. Depending upon the species, vegetative (mycelial and arthrospore) cells may be multi- or uninucleate. Chlamydospores may be formed. Ascus formation generally occurs after fusion of uninucleate gametangia, which are not differentiated by size. One to four spores are formed in the globose to ovoidal ascus. The spores, depending upon the species, may be hat shaped, spheroidal, or ovoidal with refractile bodies, and often have a gelatinous sheath. Species may possess fermentative abilities or have only an oxidative dissimilation. Surface pellicles are formed in liquid media. Nitrate is not utilized. Imperfect stage is the genus *Geotrichum.*

Schizosaccharomyces

Cells are cylindrical, elongate, ovoid, or globose. In one species a limited true mycelium is formed. The cells are haploid and reproduce by fission and disarticulation into arthrospores. Ascus formation follows conjugation between two vegetative cells (arthrospores). The zygote becomes the ascus; four to eight spheroidal, bean-shaped, or ellipsoidal spores are formed per ascus. Asci usually rupture at maturity. Members of this genus are fermentative as well as oxidative. Nitrate is not assimilated. Two of the four species of this genus form a starch-like compound in the ascospore wall, demonstrable by treatment with Lugol solution.

Found in fermented beverages, fruits, wines, and substrates of high sugar content.

Dekkera

Vegetative cells vary in shape from ovoid to ellipsoidal or elongate and are frequently ogival. Pseudomycelium is formed, although occasionally rudimentary. One to four spores are formed per ascus, which ruptures soon after maturation. Spores are hat shaped to spheroidal with a tangential brim. Dissimilation is oxidative and fermentative (slow). There is usually a vigorous production of acetic acid from glucose under aerobic conditions. Formation of surface pellicles is variable. Assimilation of nitrate is variable. This genus represents the perfect stage of the genus *Brettanomyces*. Its species have been isolated from beers and wines.

Saccharomycopsis (= Endomycopsis)

A word of explanation regarding this change in the generic name is owed those readers who recognize species of this genus used in industry by its former name *Endomycopsis*. *Saccharomycopsis* was used by Schiönning earlier (1903) to describe the type species used by Stelling-Dekker in 1931 for *Endomycopsis*. Thus, the older generic name has priority and *Endomycopsis* is illegitimate, having been perpetuated contrary to the International Code of Botanical Nomenclature that governs the naming of fungi.

Members of this genus produce abundant true mycelium with blastospores, pseudomycelium, and individual budding cells. There is occasionally disarticulation of the mycelium into arthrospores. Asci are formed terminally at the tips of hyphae or intercalarily (with or without previous conjugation) and usually rupture at maturity. The spores, one to four per ascus, vary in shape and characteristics according to species. They may be spheroidal or hat, sickle, or Saturn shaped. Members of the genus have mainly an oxidative metabolism, although, in addition, several species are weakly fermentative. Pellicle formation is often found in liquid media.

Arthroascus

Vegetative growth is a branched septate mycelium easily breaking up into hyphal cells. Budding cells are not formed, or formed on a broad base. Asci are formed from the vegetative cells (usually after conjugation) and contain one to four spheroidal ascospores with an equatorial or subequatorial ledge. Dissimilation is oxidative. Nitrate is not utilized.

Saccharomyces

Vegetative cells may be globose, ovoid, ellipsoidal, or elongate. Cells are usually in pairs or in small clusters, but a pseudomycelium may be formed in some species. True mycelium is never present. Vegetative reproduction occurs by multilateral budding. In diploid species spores are produced directly in vegetative cells. In haploid species conjugation between two vegetative cells, or between a mother cell and its bud, immediately precedes ascus formation. One to four spheroidal spores are formed in each ascus. The ascus does not rupture at maturity. All species have a strongly fermentative as well as a respiratory metabolism. Pellicles are not formed in liquid media. Nitrate is not utilized.

Kluyveromyces

Vegetative cells are ovoid to elongate, reproducing by multilateral budding. Pseudomycelium may be formed. Ascus formation is preceded by isogamic or heterogamic conjugation, or asci are derived from a diploid vegetative cell. Asci may be single to multispored (sixteen spores or more per ascus is considered multispored). Asci normally rupture at maturity. The spore shape is ovoid to kidney shaped. Fermentative as well as oxidative dissimilation occurs. A surface pellicle is formed by one species (*K. polysporus*). Nitrate is not utilized.

Schwanniomyces

Vegetative cells are spherical to ovoid in shape, reproducing by multilateral budding. Prior to ascus formation, the haploid cell forms a protuberance (meiosis bud). One to two warty-walled spores with an equatorial ledge (resembling a walnut) are formed in the mother cell. Ascus formation is rarely accomplished by conjugation between two independent vegetative cells (isogamy). Spores are not released from the ascus at maturity. These yeasts are fermentative as well as oxidative in their dissimilation. Pellicles are not formed in liquid media. Nitrate is not assimilated.

Debaryomyces

Cells are generally spheroidal to globose, reproducing by multilateral budding. Pseudomycelium is normally absent. The ascus is formed by a kind of mother–daughter cell (heterogamic) conjugation, although isogamic conjugation between two individual similar cells may occur. One or (more rarely) two warty-walled ascospores are formed. Dissimilation is oxidative in some species, but others have both a fermentative and an oxidative metabolism. Pellicle formation is variable. Nitrate is not assimilated, but some species can utilize nitrite. Most species have a high tolerance to sodium chloride.

Citeromyces

Members of this genus produce spheroidal to ovoidal vegetative cells that reproduce by multilateral budding. No pseudomycelium is formed. Asci are formed from diploid cells. One or (more rarely) two spheroidal, warty spores with a prominent oil globule are formed per ascus, which do not dehisce at maturity. Dissimilation is fermentative as well as oxidative. No pellicle is formed in liquid media. Nitrate is utilized.

Pichia

Cells range from short ellipsoidal to cylindrical, reproducing by multilateral budding. Pseudomycelium is generally formed, but may be rudimentary or lacking. Some species may form limited, true hyphae. Ascus formation occurs either directly in diploid cells or after conjugation between haploid cells. One to four Saturn-shaped, hat-shaped, or spheroidal spores are formed per ascus, and they dehisce from the ascus at maturity in most species. A dry, dull pellicle is formed in some species, but it may be thin or lacking in others. Dissimilation is preferentially oxidative, but fermentation may occur in some species. Nitrate is not utilized.

Hansenula

Vegetative cells vary from spherical or ovoid to elongate and cylindrical. Reproduction occurs by multilateral budding. Pseudomycelium is commonly formed; however, it may be rudimentary or lacking. Some species form true hyphae. Ascus formation may or may not be immediately preceded by conjugation between haploid cells. One to four spores are formed per ascus, which usually are released at maturity. Spores may be Saturn shaped to spheroidal or hat shaped. In some species a dull, dry pellicle is formed on liquid media. Some species are capable of a vigorous fermentation, although others ferment weakly or not at all. Some species produce large quantities of esters. Nitrate is utilized.

Pachysolen

Ovoid, elongate to cylindrical cells, that reproduce by multilateral budding. Pseudomycelium may be formed. A thin-walled ascus is formed at the tip of an elongated tube that develops from the mother cell. In haploid strains a conjugated small cell or bud is attached to the mother cell. During the process of tube development, the walls of the mother cell and the tube become very thick. The protoplasmic contents go into the thin-walled tip of the tube, which develops into an ascus containing four hat-shaped spores. The spores are liberated from the ascus upon maturity. Species of this genus have a weakly fermentative ability, as well as an oxidative dissimilation. Nitrate is utilized.

Cyniclomyces (Saccharomycopsis)

For an explanation of the change in generic name, refer to the description of *Saccharomycopsis*. *Cyniclomyces* was established by van der Walt and Scott to permit the proper usage of the name *Saccharomycopsis* for organisms previously placed in the illegitimate genus *Endomycopsis*.

Vegetative cells are large, elongate to cylindrical, reproducing by multilateral budding. A pseudomycelium is formed in liquid media. Vegetative growth of the single species in this genus occurs only between 30 and 40°C. An atmosphere enriched with carbon dioxide and a medium containing amino acids are required for growth. The large vegetative cells are short lived. Spore formation occurs in a diploid cell in which one to four ellipsoidal to elongate spores are formed. The ascospores have a double wall. In contrast to the temperature requirements for vegetative growth, ascospore formation occurs only at lower temperature (optimally at 18°C). *Cyniclomyces guttulatus* is weakly fermentative and does not produce a pellicle in liquid media. It can be isolated from the stomach contents, intestinal tract, or feces of domestic rabbits.

Lodderomyces

Vegetative cells are spheroidal, ellipsoidal to cylindrical, reproducing by multilateral budding. Pseudomycelium is abundantly formed. Asci, formed from diploid vegetative cells, contain one or two large oblong ascospores with obtuse ends. The spores are not liberated from the ascus at maturity. The single species has a weak fermentative ability, can utilize n–alkanes for growth, and is unable to utilize nitrate.

Wingea

The single species forms spheroidal to ellipsoidal cells that reproduce by multilateral budding. Ascus formation is usually preceded by the formation of a bud or protuberance by the haploid vegetative cell; occasionally two cells will conjugate. One to four brownish, lens-shaped ascospores are formed per ascus. Fermentative ability is rather weak, and nitrate is not utilized.

Wickerhamiella

Cells are small, globose to short ovoid, and budding is multilateral; budding on a broad base is occasionally observed. Neither true nor pseudomycelium is formed. Asci are formed after conjugation between two haploid cells. One elongate to oblong, rugose-walled ascospore is formed. The spore is liberated from the ascus at maturity. Dissimilation is oxidative, and nitrate is utilized.

Nadsonia

Vegetative cells are ellipsoidal, elongated, or sometimes lemon shaped, reproducing asexually by a process intermediate between budding and fission. Short chains of elongate cells may be found. After a heterogamic conjugation between the mother cell and a bud, the contents of the zygote move into another bud (ascus) formed at the opposite end of the mother cell. The ascus is then delimited by a septum, and one or (more rarely) two spherical, brownish, spiny-walled spores are formed. One species forms its spores within the mother cell rather than in the second bud. Dissimilation is fermentative and/or oxidative. Nitrate is not utilized.

Hanseniaspora

Cells are apiculate or ellipsoidal to elongate, reproducing vegetatively by bipolar budding. Occasionally a rudimentary pseudomycelium may be formed. Ascospores are formed directly in diploid cells. Two or four hat- to helmet-shaped or one to two spheroidal spores, each with a subequatorial ledge, are formed per ascus. Only species with helmet-shaped spores release them from the ascus upon maturity. Fermentative as well as oxidative dissimilation occurs. Pellicles are not formed. Nitrate is not utilized. Species represent the sporogenous state of members of the genus *Kloeckera*.

Saccharomycodes

Cell shape is apiculate or sausage shaped. Vegetative reproduction is by bipolar budding on a somewhat broad base. Pseudomycelium, when formed, is poorly developed.

Ascus formation occurs directly from a diploid cell. Four spheroidal spores, with indistinct subequatorial ledges, are formed per ascus. Spores conjugate in pairs within the ascus. Dissimilation is fermentative as well as oxidative. Pellicles are not formed. Nitrate is not assimilated.

Wickerhamia

The vegetative cells are apiculate, ovoid, or long ovoid. They reproduce by bipolar budding on a broad base. Cells are separated from each other by a type of bud fission. Pseudomycelium is rudimentary. The diploid vegetative cells are transformed into asci containing one or two spores per ascus. One investigator has reported that as many as sixteen spores per ascus were observed. The ascospores are cap shaped; the crown deflects to one side of a sinuous brim, giving the appearance of a sporting cap. Upon maturity the ascus splits transversely, liberating the spores. The single species of this genus is fermentative as well as oxidative in its dissimilation. A pellicle is not formed in liquid media. Nitrate is not assimilated.

Lipomyces

Generally, cells are spheroidal, globose to ovoidal with multilateral budding. Some strains, in addition, produce ovoid to cylindrical cells on a broad base and in chains. The vegetative cells are encapsulated. Pseudomycelium may be formed. Asci are formed from vegetative cells as thin-walled sac-like protuberances in which four to sixteen (occasionally more) ellipsoidal, amber-colored spores are formed. The mother cell may also contain spores in addition to those produced in the sac-like appendage. Spore walls are usually smooth, but longitudinal ridges have been observed in some strains. Asci rupture at maturity. Dissimilation is oxidative, and nitrate is not utilized. Under appropriate conditions, large fat globules are present in the vegetative cells and a starch-like compound may be formed in the capsule.

Metschnikowia

Vegetative cells range from spheroidal to cylindrical, reproduce by multilateral budding, and may form a rudimentary pseudomycelium. Chlamydospores are formed by some species. Asci originate from a vegetative cell or from a chlamydospore by elongation and are much larger than the vegetative cells. Asci are club shaped, sphaeropedunculate or ellipsoidopedunculate, and contain one or two needle-shaped spores pointed at one or both ends. Dissimilation is fermentative and/or oxidative. Nitrate is not utilized.

Nematospora

Cells are spheroidal, ovoidal, elongate, or irregular in shape (polymorphic) and reproduce by multilateral budding. A true mycelium with relatively few septa may be formed in older cultures. Pseudomycelial formations are always present. Asci are formed from a vegetative cell, which enlarges greatly prior to the formation of eight spindle- to needle-shaped ascospores. Each spore has a nonmotile, whip-like appendage. Frequently, the spores lie in two bundles of four spores each, at opposite ends of the ascus, which ruptures at maturity. Dissimilation is fermentative as well as oxidative. Nitrate is not utilized.

Ashbya

This yeast-like organism is related to *Nematospora* by the shape of its spores. Differences are that *Ashbya* has no yeast-like budding cells and vegetatively has a sparsely septate, multinucleate mycelium. The spore sacs, which are probably not true asci but sporangia, are formed within the mycelium. They contain eight to thirty-two slightly curved, needle-like spores with appendages. Metabolism is strictly oxidative. Some strains produce large amounts of riboflavin in special media.

Eremothecium (= *Crebrothecium*)

This genus is very similar to *Ashbya*, differing primarily in the shape of the spores, which are relatively short, arcuate (bowed at one end and pointed at the other), and lack an appendage. Dissimilation is strictly oxidative. Some strains produce large amounts of riboflavin in suitable media.

BASIDIOMYCETOUS YEASTS

As a group the genera included within this subdivision have budding yeast cells, mycelium with or without clamp connections, and chlamydospores. Sporidia are formed laterally or terminally on a promycelium from a germinated chlamydospore (teliospore), or terminally on a basidium. Ballistospores may be formed. Colonies may be hyaline, red, or orangeish in color. Life cycles vary among genera and species. Among these genera are representatives of the families *Ustilaginaceae* and *Tilletiaceae* (order Ustilaginales).

Leucosporidum

Cells are ovoidal to elongate, reproduce by budding, and usually have well-developed pseudomycelial growth. True mycelium may be formed. Conjugation of haploid cells of compatible mating types produces a dikaryotic, septate mycelium with clamp connections. Thick-walled chlamydospores (teliospores) with granular contents are formed intercalarily or terminally and, after karyogamy, germinate to thin-walled, one to four-celled promycelia, upon which haploid sporidia are produced laterally or terminally. A self-sporulating stage not involving conjugation of mating types produces a septate mycelium without clamp connections. The self-sporulating stage is not known for all species. Dissimilation is fermentative and/or oxidative. Nitrate is utilized. Colonies are not pigmented (hyaline). Imperfect forms are placed in the genus *Candida*.

Rhodosporidium

Yeast cells are spheroidal to elongate and reproduce by budding. Buds are often formed repeatedly at the same site. Pseudomycelium may be formed; true mycelium is rare. Haploid cells of compatible mating types conjugate, forming a dikaryotic, septate mycelium with clamp connections. The chlamydospores (teliospores) germinate to promycelia which form sporidia laterally and terminally. A self-sporulating phase similar to that which occurs in *Leucosporidium* is known for some species. Species are nonfermentative. Nitrate assimilation varies among the species. Colonies are pink to orange in color. Imperfect forms are placed in *Rhodotorula*.

Filobasidium

Cells are ovoidal to elongate and reproduce by budding. Pseudomycelium may be formed. Conjugation between haploid cells of compatible mating types gives rise to a dikaryotic, septate mycelium with clamp connections. Ultimately, a long slender basidium is formed that is nonseptate. At a somewhat swollen tip five to eight thin-walled basidiospores are formed, which upon germination produce blastospores. The basidiospores are not forcibly discharged from the basidium. Dissimilation is fermentative and/or oxidative. Assimilation of nitrate is variable.

Filobasidiella

Cells are spheroidal to elongate and reproduce by budding. Conjugation between haploid cells of compatible mating types creates a septate mycelium with clamp connections. Budding cells may arise from this mycelium, which also produces the basidia. Slender, nonseptate basidia that are swollen at the apices produce terminal, small, thin-walled basidiospores in chains. The formation of basidiospores in chains distinguishes this genus from *Filobasidium.* It was established to accomodate the perfect stage of *Cryptococcus neoformans.*

Aessosporon

Cells are globose, ovoidal to elongate, reproducing by budding and by the formation of symmetrical or asymmetrical (reniform, falcate, subglobose, or ellipsoidal) ballistospores. Pseudo- and true mycelium may be formed. Thick-walled chlamydospores (teliospores) germinate to nonseptate or one-septate promycelia, upon which one to four budding, sessile sporida are formed. Dikaryotic mycelium is not formed. Teliospores apparently arise from diploid cells. Dissimilation is oxidative. Nitrate assimilation varies among the species. Carotenoid pigments may be formed. Imperfect strains are placed in *Bullera* or in *Sporobolomyces.*

Sporidiobolus

Ovoidal to elongate yeast cells reproducing by budding and by the formation of asymmetrical ballistospores. A dikaryotic, septate mycelium with clamp connections apparently arises from a diploid yeast cell following reduction division. Chlamydospores, which become uninucleate, form on the dikaryotic mycelium. Upon germination the chlamydospores produce uninucleate, diploid yeast cells. The species are nonfermentative and assimilate nitrate.

ASPOROGENOUS YEASTS

BALLISTOSPOROGENOUS GENERA

Sporobolomyces

Cells are ovoidal to elongate, reproducing by budding. True mycelium is formed in some species. Pseudomycelium may also be found. Ballistospores are generally kidney to sickle shaped (asymmetric in form) and are produced on aerial sterigmata. They are forcibly discharged. Dissimilation is oxidative. Nearly all species appear red to salmon-pink due to the production of carotenoid pigments. In liquid media, mainly surface growth occurs. Nitrate utilization varies among the species.

Bullera

Cells are ovoidal to spheroidal in shape and reproduce by budding. True and pseudomycelium are absent. The ballistospores are typically symmetrical, being spheroidal to ovoidal in shape. They are forcibly discharged from aerial sterigmata. Dissimilation is strictly oxidative. In culture these organisms are colorless to pale yellow or cream colored. In liquid media principally surface growth occurs. Nitrate is not utilized.

GENERA WITHOUT BALLISTOSPORES

Cryptococcus

Cells are spheroidal, ovoidal, occasionally elongate, or irregularly shaped. Reproduction occurs by budding. Pseudomycelium is lacking. Cells of most species are surrounded by a capsule, giving colonies a mucoid appearance. Growth on solid media may be pallid or somewhat yellowish, pinkish, or red due to the synthesis of carotenoid pigments. Starch-like compounds are formed by most species of this genus. Nitrate utilization is variable. Inositol or glucuronic acid is assimilated.

Rhodotorula

Cells are spheroidal, ellipsoidal, or elongate reproducing by budding. A more or less rudimentary pseudo- or true mycelium may occasionally be formed. Many species have a mucoid appearance due to formation of a capsule. Dissimilation is strictly oxidative. Distinct red or yellow carotenoid pigments are produced. Dry surface pellicles are not formed. Nitrate utilization varies with the species. Inositol is not assimilated.

Phaffia

Vegetative cells are mainly ellipsoidal, and they reproduce by budding. A rudimentary pseudomycelium may be present. Chlamydospores are formed. Fermentative ability is present, and starch-like compounds are synthesized. Nitrate is not utilized. Colonies are pigmented due to astaxanthin formation (with a minor proportion of β–carotene), giving them a salmon-pink color.

Pityrosporum

The shape of the vegetative cell is characteristically flask-like or ovoidal. Vegetative reproduction occurs by budding of ellipsoidal cells at one of the poles on a very broad base (bud fission), giving the vegetative cells a bottle-like shape. Pseudomycelium or true mycelium are very rarely formed. Dissimilation is strictly oxidative. Species of this genus do not grow well on malt extract media, but growth is stimulated by the addition of certain higher fatty acids. Habitat is on the skin of warm-blooded animals, and optimum temperature for growth is about 36°C.

Schizoblastosporion

Cells are ellipsoidal to elongate. Reproduction takes place by a combination of budding and fission in a bipolar manner. A pseudomycelium is not formed. Metabolism is strictly oxidative. Pellicle formation in liquid media varies between strains of this monotypic genus. Nitrate is not utilized.

Kloeckera

The vegetative cell shape is apiculate or ellipsoidal. Reproduction is by budding at the poles. Occasionally, a rudimentary pseudomycelium is formed. Fermentative as well as oxidative dissimilation occurs. Pellicles are absent. Nitrate is not assimilated. Its perfect state is *Hanseniaspora.*

Trigonopsis

Cells are typically triangular or ellipsoidal in shape. In the triangular cells budding occurs at the three apices of the cells; in ellipsoidal cells, multilaterally. No pseudomycelium or true mycelium is formed. The single species of this genus is strictly oxidative. A thin surface pellicle may be formed. Nitrate is not utilized.

Brettanomyces

Vegetative cells are often ogival, ovoidal, or spheroidal; elongated cells also occur. Vegetative reproduction is by budding, often resulting in irregular chains of cells. Occasionally a rudimentary pseudomycelium is formed. Fermentative as well as oxidative dissimilation occurs.

These organisms characteristically produce considerable amounts of acetic acid under aerobic conditions. Fermentation of malt extract is slow and gives rise to a characteristic aroma. Pellicle formation and nitrate assimilation are variable. Its perfect state is *Dekkera*.

Torulopsis

Cells are generally spheroidal to ellipsoidal, or infrequently somewhat elongated. Vegetative reproduction occurs by budding. Pseudomycelium is not formed, although rarely a rudimentary structure may develop. Fermentative dissimilation is present in most species, but it is absent in some. A few species are mucoid due to the formation of capsules. The formation of starch-like compounds or the synthesis of red or yellow carotenoid pigments does not occur. Nitrate utilization is variable. *Torulopsis* represents various perfect genera.

Candida

Vegetative cells range from spheroidal to cylindrical, reproducing by budding. Pseudomycelium is more or less abundantly formed, and in addition true mycelium may occur. The latter does not disarticulate into arthrospores. In certain species chlamydospores may be formed. Fermentative dissimilation may be strong, weak, or absent, depending on the species. Pellicle formation and nitrate assimilation are variable. *Candida* represents various perfect genera.

Trichosporon

Species are characterized by abundant development of pseudomycelium and true mycelium. The latter breaks up into arthrospores. Blastospores on the pseudomycelium or on the mycelium reproduce vegetatively by budding. Endospores and/or chlamydospores are formed by certain species. Dissimilation is mainly oxidative, although some species are weakly fermentative. Thick surface pellicles are formed. Nitrate utilization varies with the species.

Oosporidium

Vegetative cells are of various shapes, reproducing by budding on a broad base (bud fission), and they commonly appear in chains. Occasionally true septa are formed but arthrospores are lacking. Endospores are formed intracellularly by protoplasmic cleavage. Colonies are pink to orange-yellow, but the pigments are not carotenoid in nature. Dissimilation is oxidative, and nitrate is assimilated.

Sterigmatomyces

Cells are spheroidal to ovoidal, reproducing by the production of sterigmata or conidiophore-like protuberances with a conidium developing at the terminus. The single conidial cell is separated from the mother cell by a septum-like wall in the midregion of the "stalk". This form of reproduction is then repeated with or without cell separation. No true budding cells, pseudo- , or true hyphae are formed, but chains of cells occur in some species. Dissimilation is oxidative, and utilization of nitrate varies with the species. Colonies are nonpigmented.

Sympodiomyces

Cells are ovoidal and reproduce by a sympodial mechanism where a conidiophore develops directly from a yeast cell, producing a terminal conidium. A new growing tip of another condiophore develops adjacent to the first, forming a new conidium. A rudimentary true mycelium may develop. Dissimilation is oxidative, and nitrate is not assimilated.

Members of this and the preceding genus are not strictly "budding" yeasts, but are unicellular in their vegetative stage. They have been isolated from marine waters.

RELATED YEAST-LIKE ORGANISMS

Aureobasidium (syn. *Pullularia* and *Dematium*)

Members of this genus are commonly known as "black yeasts." They are found on plant materials and in soil and are frequently isolated with yeast. Most authorities recognize but a single species *(A. pullulans)* because of the numerous variations caused by genetic instability. The appearance of a young colony is from white or tan to light pink; upon aging, colonies turn greenish and finally black. Some strains, however, do not complete this color transition. Colonies vary in texture from pasty or mucoidal to tough and leathery. The edge of the colony is strongly rhizoidal in appearance. Vegetative reproduction varies from mainly yeast-like (multilateral budding) to strongly mycelial. Arthrospores, chlamydospores, and blastoconidia are found commonly. Young vegetative cells are ellipsoidal to apiculate. Metabolism is oxidative.

Geotrichum (Syn. *Oidium, Oospora)*

Members of this genus grow vegetatively as mycelium, which readily breaks up into arthrospores. Microscopic observation of such cultures frequently shows virtually nothing but arthrospores and very few mycelial strands. Since this is a common fungus of worldwide distribution and varied habitat, it frequently appears on yeast isolation plates. The colonies are yeast-like when young; they develop rapidly into white, spreading colonies with chains of arthrospores, and show dichotomous branching of the hyphae at the margin. A very weak fermentative ability may be observed, but the dissimilation is principally oxidative. Cultures appear to be genetically unstable. Perfect states are included in the genus *Endomyces*.

Taphrina

Species of this genus are parasitic molds on plants, causing a condition known as curly leaf. In culture they exist as multilaterally budding yeast cells. Colonies are pasty, glistening, and often slightly pinkish in color. For these reasons *Taphrina* may easily be confused with a typical yeast. Dissimilation is oxidative, but surface pellicles are not formed.

Prototheca

Members of this genus are colorless algae, probably derived from the green alga *Chlorella*. The colorless, spherical to ellipsoidal cells may be mistaken for yeast cells. However, budding does not occur. Vegetative reproduction in *Prototheca* is by a partitioning of the cellular protoplasm into spore-like bodies (spherules or aplanospores). A mother cell with spherules resembles a yeast ascus superficially.

Appendix B / Principal Characteristics of the Yeast Genera Listed in the Text.

The names are listed alphabetically for rapid orientation. The utilization of nitrate is indicated only if positive or variable, depending on the species. Names followed by an asterisk are "related" yeast-like genera.

Aessosporon — Cells are globose, ovoidal to elongate reproducing by multilateral budding and by formation of symmetrical or asymmetrical (reniform or falcate) ballistospores. Pseudo- and true mycelium may be formed. Thick-walled chlamydospores (teliospores) germinate to a nonseptate promycelium, upon which one to four budding sporidia are formed. Dikaryotic mycelium is not formed. Teliospores apparently arise from diploid cells. Dissimilation is oxidative; nitrate assimilation is variable. Carotenoid pigments may be formed. This genus represents the perfect state of both *Bullera* and *Sporobolomyces*.

Arthroascus — Vegetative growth is a branched septate mycelium easily breaking up into hyphal cells. Budding cells are not formed, or formed on a broad base. Asci originate from the vegetative cells (sometimes after conjugation) and contain one to four hat- to Saturn-shaped ascospores. Dissimilation is oxidative.

*Ashbya** — Sparsely septate, multinucleate mycelium; eight to thirty-two needle-shaped ascospores with appendages occurring in groups in the mycelium; fermentation absent; some strains form large amounts of riboflavin.

*Aureobasidium** — Budding cells, true mycelium, pseudomycelium, arthrospores; no ascospores; no fermentation; most strains become dark brown or black in culture.

Brettanomyces — Cells ovoidal and ogival (pointed at one end), vegetative reproduction by budding, ascospores not formed; fermentation slow; a characteristic aroma is produced; aerobically, alcohol is oxidized to acetic acid; nitrate assimilation variable.

Bullera — Budding, ovoidal cells; symmetrical ballistospores are forcibly discharged; fermentation absent. Sexual spores not formed.

Candida — Pseudomycelium, budding cells, sometimes true mycelium; arthrospores absent; sexual spores not formed; fermentation and nitrate assimilation variable.

Citeromyces — Ovoidal, budding cells; one or two warty, spherical spores per ascus; fermentation positive; nitrate is assimilated.

Coccidiascus — Ovoidal, multilaterally budding cells; four needle-shaped spores (without appendages) per ascus. Has been observed in *Drosophila*, but has not been cultivated.

Cryptococcus — Cells are spherical, ovoidal, occasionally elongate, or irregularly shaped. Reproduction occurs by budding. Pseudomycelium is lacking. Cells of some species are surrounded by a capsule, giving colonies a mucoid ap-

	pearance. Growth on solid media may be pallid to somewhat yellowish or pinkish, due to the synthesis of small amounts of carotenoid pigments. Starch-like compounds are formed on acidic media by most species. Fermentative ability is lacking. Dry surface pellicles are not formed. Nitrate utilization is variable. Inositol and (or) D–glucuronic acid are assimilated.
Cyniclomyces (= Saccharomycopsis)	Budding cells large, elongate; one to four ellipsoidal ascospores per ascus; the single species requires CO_2, amino acids, and 30–40°C for growth; fermentation weak.
Debaryomyces	Spheroidal, multilaterally budding cells; one to four warty spores per ascus; fermentation variable; often a high tolerance to NaCl.
Dekkera	This genus is the sporogenous equivalent of *Brettanomyces;* asci contain one to four hat-shaped or spherical spores, the latter with a tangential brim.
*Endomyces**	Septate mycelium and arthrospores; asci usually borne from conjugating gametangial hyphal tips, and they contain one to four spores, which are hat shaped or spherical with a smooth or

	wrinkled membrane; fermentation variable.
*Eremascus**	True mycelium with cross walls, but no arthrospores; mold-like; eight ovoidal ascospores per ascus; fermentation absent; requires high sugar levels.
*Eremothecium** (= *Crebrothecium*)	Septate mycelium; groups of numerous arcuate ascospores are found in the mycelium; fermentation absent; riboflavin is formed abundantly.
Filobasidiella	Cells are spherical to globose and reproduce by budding. Conjugation between haploid cells of compatible mating types gives rise to a dikaryotic, septate mycelium with clamp connections. Long basidia, each having a subglobose to flask-shaped apex with basipetal chains of basidiospores formed by repetitious budding. Teliospores are lacking. Basidiospores upon germination produce budding cells. Dissimilation is oxidative and nitrate is not utilized. The imperfect form is *Cryptococcus neoformans*.
Filobasidium	Cells are ovoidal to elongate and reproduce by budding. Pseudomycelium may be formed. Conjugation between haploid cells of compatible mating types gives rise to a dikaryotic, septate

	mycelium with clamp connections. Ultimately a long slender basidium is formed that is nonseptate. At a somewhat swollen tip five to eight thin-walled basidiospores are formed, which upon germination produce blastospores. Dissimilation is fermentative and/or oxidative. Assimilation of nitrate is variable.
*Geotrichum**	Septate mycelium and arthrospores; mold-like; no ascospores are formed; spreading, white, powdery growth; fermentation absent or very weak. It is the arthroconidial state of *Endomyces*.
Hanseniaspora	Lemon-shaped or ovoidal cells; bipolar budding; asci contain either two to four hat-shaped spores or one to two spherical spores with indistinct equatorial ledges; fermentation positive.
Hansenula	Budding cells of various shapes, sometimes pseudomycelium, and/or true mycelium; two to four hat-shaped or Saturn-shaped spores per ascus; fermentation variable; nitrate is assimilated.
Kloeckera	Lemon-shaped or ovoidal cells; bipolar budding; ascospores not formed; fermentation positive; imperfect form of *Hanseniaspora*.

Kluyveromyces — Multilaterally budding ovoidal cells; asci usually one to four spored; some species multispored; spores kidney shaped to ovoid and asci rupture at maturity; fermentation positive.

Leucosporidium — Cell are ovoidal to elongate, reproduce by budding, and usually have well-developed pseudomycelial growth. True mycelium may be formed. Conjugation of haploid cells of compatible mating types produces a dikaryotic, septate mycelium with clamp connections. Thick-walled teliospores with granular contents are formed intercalarily or terminally and, after karyogamy, germinate to thin-walled, one- to four-celled promycelia, upon which haploid sporidia are produced laterally or terminally. In some species there may also be a self-sporulating stage not involving conjugation of mating types; it produces a septate mycelium without clamp connections. Dissimilation is fermentative and/or oxidative. Nitrate is utilized. Colonies are not pigmented (hyaline). Imperfect forms are placed in the genus *Candida*.

Lipomyces — Capsulated, multilaterally budding cells of various shapes; four to sixteen ovoid, amber-colored spores per ascus; fermentation absent; habitat soil.

Lodderomyces — Vegetative cells are spheroidal to cylindrical, reproducing by multilateral budding. Pseudomycelium may be formed. Diploid vegetative cells are converted to asci, which contain one or two large oblong ascospores with obtuse ends. The spores are not liberated at ascus maturity. The single species has a weak fermentative ability.

Metschnikowia — Multilaterally budding, ovoidal cells; chlamydospores may be formed; asci large, club shaped, containing one or two needle-shaped spores without appendages; fermentation variable, depending on the species.

Nadsonia — Cells ovoidal, lemon shaped, or elongate; bipolar budding on a broad base (bud fission); one or two brown, spherical, spiny spores per ascus; fermentation present.

Nematospora — Cells polymorphic; multilateral budding; sometimes true mycelium; asci large, containing eight needle-shaped spores with whip-like appendages; fermentation positive; plant parasites.

Oosporidium — Vegetative cells of various shapes, reproducing by multilateral budding on a broad base (bud fission) and commonly appearing as cell chains. Occasionally true septa are formed, but no arthrospores. Asexual endospores are formed intracellularly by protoplasmic cleavage. Colonies are pink to orange-yellow, but pigments are not carotenoid in nature. Dissimilation is oxidative; nitrate is assimilated.

Pachysolen Ovoidal to elongate budding cells, surrounded by a capsule; asci are borne at the tips of ascophores, thick-walled special structures; there are four hat-shaped spores per ascus, which ruptures at maturity; fermentation is weak; nitrate is assimilated.

Phaffia Vegetative cells are mainly ellipsoidal and reproduce by budding; they may form a rudimentary pseudomycelium. Chlamydospores are formed. Fermentative ability is present, and starch-like compounds are formed. Colonies are salmon-pink; the pigments consist mainly of astaxanthin, and there is a minor proportion of β–carotene.

Pichia Ovoidal to elongate multilaterally budding cells; pseudomycelium variable; rarely true mycelium; two to four helmet-, hat-shaped, or spherical spores per ascus; fermentation variable.

Pityrosporum Cells oval or flaskshaped, reproducing by a combination of budding and fission at one of the poles of the cell; no ascospores; no fermentation; requires lipids for growth.

*Prototheca** — Oval to spherical cells, which multiply by internal partitioning, forming two to numerous progeny cells or aplanospores; presumed to be colorless algae derived from *Chlorella*.

Rhodosporidium — Yeast cells are spheroidal to elongate and reproduce by budding. Pseudomycelium may be formed, true mycelium is rare. Haploid cells of compatible mating types conjugate, forming a dikaryotic, septate mycelium with clamp connections. The chlamydospores (teliospores) germinate to septate or nonseptate promycelia, which form sporidia laterally or terminally. A self-sporulating phase as in *Leucosporidium* is known for some species. Species are nonfermentative; nitrate assimilation is variable. Cell growth is pink to orange in color due to formation of carotenoid pigments. Imperfect forms are placed in *Rhodotorula* or *Cryptococcus*.

Rhodotorula — Budding cells, usually ovoidal to elongate; sexual spores absent; fermentation absent; nitrate assimilation variable; pink carotenoid pigments present.

Saccharomyces	Multilaterally budding cells; one to four ovoidal or spherical spores per ascus; ascus wall does not lyse upon maturity; fermentation strong.
Saccharomycodes	Cells lemon shaped, bipolar budding on a broad base; asci contain four spherical spores that may contain a narrow ledge; the single species is fermentative.
Saccharomycopsis (= *Endomycopsis*)	Well developed true mycelium, pseudomycelium, budding cells, rarely arthrospores; ascospores (one to four per ascus) spherical, ellipsoidal, kidney shaped, hat shaped or Saturn shaped; fermentation variable, but weak if present.
Schizosaccharomyces	Cells elongate, reproducing by cross-wall formation; four or eight oval to kidney-shaped ascospores per ascus; rudimentary mycelium in one species; fermentation positive.
Schizoblastosporion	Cells ovoidal or flask shaped, reproducing by a combination of budding and fission; no ascospores formed; fermentation absent; no requirement for lipids.
Schwanniomyces	Spheroidal budding cells; one or two walnut-shaped ascospores per ascus;

	meiosis buds often present; fermentation positive.
Sporidiobolus	Ovoidal to elongated vegetative cells reproducing by budding and by the formation of asymmetrical ballistospores. A dikaryotic, septate mycelium with clamp connections arises from a diploid yeast cell. Chlamydospores form on the dikaryotic mycelium. These spores apparently germinate into diploid yeast cells. The haplophase is believed to be the result of reduction within a vegetative cell prior to the formation of the dikaryotic mycelium, from which ballistospores may be formed. Strains are nonfermentative; nitrate is assimilated.
Sporobolomyces	Budding cells, sometimes true mycelium and pseudomycelium; asymmetrical ballistospores are forcibly discharged; fermentation absent; nitrate assimilation variable. Asexual state of *Aessosporon*.
Sterigmatomyces	Cells spheroidal to ovoidal reproducing by the production of sterigmata or conidiophore-like protuberances with a conidium developing at the terminus and separating from the mother cell by a septum-like wall in the midregion of the "stalk." No budding cells, pseudo- or true hyphae are formed. Dissimila-

tion is oxidative; utilization of nitrate is variable among the species. Colonies are nonpigmented.

Sympodiomyces Cells are ovoidal and reproduce by a sympodial mechanism where a conidiophore develops directly from a yeast cell, producing a terminal conidium. A new growing tip of another conidiophore develops adjacent to the former, forming a new conidium. A rudimentary true mycelium may develop. Dissimilation is oxidative. Members of this and the preceding genus are not strictly "budding" yeasts but are unicellular in their vegetative stage. They have been isolated from marine waters.

*Taphrina** Multilaterally budding cells on artificial media; parasitic on plants, where mycelial growth occurs, producing eight-spored asci; no fermentation.

Torulopsis Cells reproduce by budding; sexual spores not formed; pseudomycelium absent; fermentation variable; nitrate assimilation variable; starch-like compounds not formed; inositol and D–glucuronic acid are not assimilated.

Trichosporon True mycelium, arthrospores, pseudomycelium, and budding cells; no sexual spores formed; fermentation absent or weak; nitrate assimilation variable.

Trigonopsis — Cells triangular (budding at the apices) or ellipsoidal; no ascospores formed; the single species does not ferment.

Wickerhamia — Cells lemon shaped or ovoidal, bipolar budding on a broad base; one to two cap-shaped ascospores per ascus; the single species is fermentative.

Wickerhamiella — Cells are small, globose to short ovoidal, and budding is multilateral; budding on a broad base is occasionally observed. Neither true mycelium nor pseudomycelium is formed. Asci are formed after conjugation between two haploid cells and produce one elongate to oblong, rugose-walled ascospore. The spore is liberated at ascus maturity. Dissimilation is oxidative; nitrate is utilized.

Wingea — The single species forms spheroidal to ellipsoidal cells that reproduce by multilateral budding. Ascus formation is usually preceded by the formation of a bud or protuberance by the haploid vegetative cell; occasionally two cells conjugate. One to four brownish, lens-shaped ascospores are formed per ascus. Fermentative ability is weak.

Glossary of Certain Mycological Terms Used in the Text

Apiculate: A cell shape, somewhat resembling the form of a lemon, having protuberances at both ends of the long axis of the cell.

Arthrospore: A nonsexual spore resulting from the disarticulation of hyphae or of single cells dividing by cross-wall formation. Sometimes called oidium.

Ascospore: A sexual spore borne in an ascus.

Ascus (pl. asci): A sac-like structure containing the ascospores formed by certain yeasts.

Ballistospore: An asexual spore borne on a sterigma and forcibly discharged.

Basidiospore: A sexual spore borne on the outside of a basidium, usually on a specialized structure of pointed shape called a sterigma.

Basidium(pl. basidia): A structure (typical of the subdivision *Basidiomycotina*) in which nuclear fusion and reduction division occur, and which bears the externally located basidiospores.

Basipetal: The direction of growth from the base, with the apical portion oldest.

Blastospore: A nonsexual reproductive cell formed by budding, normally near the apex of an elongated cell, or in the area where pseudomycelial cells are joined together.

Blastoconidium: An individual nonsexual reproductive cell formed by budding along a hyphal filament, but not located at the apex where elongated cells of a pseudomycelium are joined together or at the septa of true hyphae.

Chlamydospore: A nonsexual resting cell, enveloped by a thick cell wall.

Cisterna: A sac-like structure formed by the endoplasmic reticulum.

Clamp connection: A bridge-like connection between cells of a binucleate mycelium of basidiomycetous yeasts. It ensures an orderly division of the two nuclei of opposite sex from one cell to the next (Fig. 19).

Conjugation: The fusion of two individual cells (gametes), usually followed by nuclear fusion and reduction division.

Conjugation tube: A tube-like protuberance usually formed by each of the two gametes participating in the conjugation process. The tips of the conjugation tubes (which may be very short or long) fuse and grow together.

Dangeardium: The structure in which karyogamy and meiosis take place.

Dikaryotic: The binucleate condition; usually refers to a cell containing a pair of nuclei, each derived from a different parent.

Diploid: Containing twice the basic number of chromosomes($2n$).

Diplophase: That part of the life cycle which represents the diploid condition.

Dolipore septum: A septum with a central pore surrounded by the swollen edge of the cross wall; pores may be plugged or capped on both ends (Fig. 10a).

Endospore: An asexual spore formed inside a vegetative cell.

Epiplasm: The residual protoplasm in an ascus after ascospore formation.

Fimbria: A hair-like structure or fibril on the cell surface.

Fission: An asexual reproductive process, in which a cell forms a cross wall (septum) and is then separated into two cells along the cross wall. This process is exemplified by members of the genus *Schizosaccharomyces*—the "fission yeasts."

Gametangium (pl. gametangia): A structure that contains gametes.

Gamete: A differentiated sex cell, or sex nucleus, that normally fuses with another in sexual reproduction.

Genus (pl. genera): A taxonomic category, or taxon, that is composed of one or more species. The generic name is the first word in a binomial designation of a species.

Haploid: Containing the basic (or reduced) number of chromosomes ($1n$).

Haplophase: That part of the life cycle which represents the haploid condition.

Heterogamous (also heterogamic): Usually refers to morphologically different gametes; male and female gametes in filamentous fungi.

Heterothallic: Fungi (or yeasts) in which the sexes are separated in separate thalli; thus, gametes from two different thalli are required for sexual reproduction.

Heterozygous: The diploid condition in which the nucleus is the result of the union of two dissimilar haploid nuclei.

Homothallic: Yeasts in which diploidization and sexual reproduction can take place in a culture derived from a single ascospore or haploid cell.

Homozygous: A character or trait of a diploid yeast in which both of the contributing nuclei contain identical genes for that particular character.

Hypha (pl. hyphae): One of the tube-like or thread-like elements that make up the mycelial structure of a yeast or fungus.

Intercalary: Refers to a position (usually of a spore or ascus) within a hypha (as opposed to terminal).

Karyogamy: The fusion of two nuclei.

Meiosis: Reduction division of a nucleus during sexual reproduction (formation of sexual spores). Reduction refers to a decrease in number of chromosomes per nucleus—for example, from $2n$ to $1n$.

Mitosis: Normal nuclear division with retention of the original number of chromosomes.

Mycelium (pl. mycelia—Also termed true mycelium in the literature on yeasts): A mass of true hyphae, usually with cross walls, constituting the vegetative body or thallus of a yeast or filamentous fungus.

Ogival: A cell shape where one end of an elongated cell is pointed and the other end rounded.

Parasexuality: A process in which plasmogamy, karyogamy, and haploidization take place in sequence, but not at specified points in the life cycle of an individual. It is of significance in heterokaryotic individuals that derive some of the benefits of sexuality from a parasexual cycle.

Parthenogenesis: The development of the normal product of sexual reproduction from the female gamete alone. The apomictic development of haploid cells.

Plasmalemma: A cytoplasmic membrane.

Plasmodesma (pl. plasmodesmata): A protoplasmic thread through a septum that provides protoplasmic continuity in a hypha (Fig. 10b).

Pellicle: A skin-like or film-like surface growth of a yeast on liquid media.

Ploidy: Refers to the number of basic sets of chromosomes ($1n$, $2n$, $3n$, and so on).

Promycelium: A thin-walled, tube-like outgrowth from a germinating teliospore, where meiosis of the diploid nucleus takes place. The promycelium may be septate or nonseptate.

Pseudomycelium: A series of cells that remain attached to each other, forming branched chains. In contrast to true mycelium, the component cells originally arose by budding rather than by cross-wall formation. The component hyphae are usually termed pseudohyphae.

Septate: With cross walls.

Septum (pl. septa): A cross wall in a hypha.

Species (sing. and pl. species): Normally the lowest unit of classification. A group of related individuals that compose a genus. Species are designated by binomials consisting of the generic name (first word) followed by the specific epithet.

Sporidium: A thin-walled sexual spore formed by budding on the promycelium. The (usually four) sporidia contain the haploid nuclei formed during meiosis in the promycelium.

Sterigma (pl. sterigmata): A small stalk-like structure that supports a spore. In yeasts it is the structure supporting the ballistospore formed by species of *Sporobolomyces*, *Bullera*, and related perfect genera.

Teliospore: A thickwalled structure found in certain basidiomycetous yeasts where karyogamy takes place. Teliospores occur terminally or intercalarily on dikaryotic mycelium.

Thallus (pl. thalli): The vegetative body (somatic phase) of a yeast or fungus.

Zygote: The product of nuclear fusion. Also, the cell or fusion product of two cells in which karyogamy has taken place (often dumbbell-shaped).

Selected Bibliography

Books

(Books marked with an asterisk contain chapters by various authors covering aspects of the general subject matter of this volume.)

Ainsworth, G. C., and K. Sampson. 1950. *The British Smut Fungi (Ustilaginales).* Commonwealth Mycological Institute, Kew, Surrey, England.

Alexopoulos, C. J. 1962. *Introductory Mycology*, Art work by Sung Huang Sun. 2nd ed. John Wiley, New York.

Barnett, J. A., and R. J. Pankhurst. 1974. *A New Key to the Yeasts*—A Key for Identifying Yeasts Based on Physiological Tests Only. North-Holland Publishing Co., Amsterdam.

Brock, T. D. 1961. *Milestones in Microbiology.* Prentice-Hall, Englewood Cliffs, N.J.

Condit, I. J. 1947. *The Fig*. Chronica Botanica, Waltham, Mass.

Dobell, C. 1932. *Antonie van Leeuwenhoek and His "Little Animals."* Staples Press, London.

Fischer, G. W., and C. S. Holton. 1957. *Biology and Control of the Smut Fungi.* Ronald Press, New York.

Guilliermond, A. 1920. *The Yeasts* (trans. F. W. Tanner). John Wiley, New York.

———. 1928. *Clef dichotomique pour la détermination des levures.* Librairie le François, Paris.

Harden, A. 1932. *Alcoholic Fermentation,* 4th ed. Longmans, Green, London.

Kudriavtsev, V. I. 1960. *Die Systematik der Hefen.* Akademie-Verlag, Berlin (originally published in Moscow in Russian, 1954).

Lindegren. C. C. 1949. *The Yeast Cell, Its Genetics and Cytology.* Educational Publishers, St. Louis, Mo.

Lodder, J., and N. J. W. Kreger-van Rij. 1952. *The Yeasts — A Taxonomic Study.* North-Holland Publishing Co., Amsterdam.

Reed, G., and H. J. Peppler. 1973. *Yeast Technology.* Avi Publishing Co., Westport, Conn.

Rippon, J. 1974. Medical Mycology—The Pathogenic Fungi and the Pathogenic Actinomycetes. W. B. Saunders, Philadelphia.

* *Aspects of Yeast Metabolism.* 1968. A. K. Mills, ed. F. A. Davis, Philadelphia.

* *Biochemistry of Industrial Micro-organisms.* 1963. C. Rainbow and A. H. Rose, eds. Academic Press, New York.

* *Conversion and Manufacture of Foodstuffs by Microorganisms.* 1971. Sixth International Symposium of International Union of Food Science and Technology, Kyoto, Japan. Saikon Publishing Co., Tokyo.

* *Die Hefen*, vol. 1: *Die Hefen in der Wissenschaft*. 1960. F. Reiff, R. Kautzmann, H. Lüers, and M. Lindemann, eds. Verlag Hans Carl, Nuremberg, Germany.

* *Die Hefen*, vol. 2: *Technologie der Hefen*. 1962. F. Reiff, R. Kautzmann, H. Lüers, and M. Lindemann, eds. Verlag Hans Carl, Nuremberg, Germany.

* *Fermentation Technology Today*. 1972. Proceedings of the IVth International Fermentation Symposium. G. Terui, ed. Published by the Society of Fermentation Technology, Osaka, Japan.

* *Industrial Fermentations*, vols. 1 and 2. 1954. L. A. Underkofler and R. J. Hickey, eds. Chemical Publishing Co., New York.

* *Methods in Cell Biology*. 1975. D. M. Prescott, ed. vols. 11 and 12: Yeast Cells. Academic Press, New York.

* *Microbial Growth on* C_1-Compounds. 1975. Proceedings of the International Symposium on Microbial Growth on C_1-Compounds, Tokyo. Edited by the Organizing Committee. Published by the Society of Fermentation Technology, Japan.

* *Recent Trends in Yeast Research*. 1970. D. G. Ahearn, ed. Spectrum. Monograph Series in the Arts and Sciences. Georgia State University, Atlanta, Ga.

* *The Chemistry and Biology of Yeasts*. 1957. A. H. Cook, ed. Academic Press, New York.

* *The Yeasts*—A Taxonomic Study. 1970. J. Lodder, ed. North-Holland Publishing Co., Amsterdam.

* *The Yeasts*, A. H. Rose and J. S. Harrison, eds. Academic Press, London and New York. Vol. 1 *Biology of Yeasts*, 1969; Vol. 2. *Physiology and Biochemistry of Yeasts*, 1971; Vol. 3. *Yeast Technology*, 1970.

* *Yeast, Mould and Plant Protoplasts*, 1973, J. R. Villanueva, I. Garcia-Acha, S. Gascon, and F. Uruburu, eds. Academic Press, London and New York.

* *Microbial and Plant Protoplasts*. 1976. J. F. Peberdy, A. H. Rose, H. J. Rogers, and E. C. Cocking, eds. Academic Press, London and New York.

Reviews and Other Selected Publications

Ainsworth, G. C. 1973. "Introduction and Keys to Higher Taxa." In *The Fungi*, vol. 4A. G. C. Ainsworth, F. K. Sparrow, and A. S. Sussman, eds. Academic Press, New York and London.

Ballou, C. E. 1974. "Some Aspects of the Structure, Immunochemistry, and Genetic Control of Yeast Mannans." *Adv. Enzymol. 40*, 239–270.

Barnett, J. A. 1968. "Biochemical Differentiation of Taxa with Special Reference to the Yeasts." In *The Fungi*, vol. 3. G. C. Ainsworth and A. S. Sussman, eds. Academic Press, New York and London.

Beran, K. 1968. "Budding of Yeast Cells, Their Scars and Ageing." *Adv. Microbial Physiol. 2*, 143–171.

Cabib, E. 1975. "Molecular Aspects of Yeast Morphogenesis." *Ann. Rev. Microbiol. 29*, 191–214.

Campbell, I. 1974. "Methods of Numerical Taxonomy for Various Genera of Yeasts." *Adv. Appl. Microbiol. 17*, 135–156.

Fell, J. W. 1974. "Heterobasidiomycetous Yeasts *Leucosporidium* and *Rhodosporidium*. Their Systematics and Sexual Incompatibility Systems." *Trans. Mycol. Soc. Japan 15*, 316–323.

Gorin, P. A. J., and J. F. T. Spencer. 1970. "Proton Magnetic Resonance Spectroscopy—An Aid in Identification and Chemotaxonomy of Yeasts." *Adv. Appl. Microbiol. 13*, 25–89.

Hartwell, L. H. 1974. "*Saccharomyces cerevisiae* Cell Cycle." *Bacteriol. Rev. 38*, 164–198.

Kreger-van Rij, N. J. W. 1973. "Endomycetales, Basidiomycetous Yeasts, and Related Fungi." In *The Fungi*, vol. 4A, G. C. Ainsworth, F. K. Sparrow, and A. S. Sussman, eds. Academic Press, New York and London.

Kreger-van Rij, N. J. W., and M. Veenhuis. 1971. "A Comparative Study of the Cell Wall Structure of Basidiomycetous and Related Yeasts." *J. gen. Microbiol. 68*, 87–95.

———. 1975. "Electron Microscopy of Ascus Formation in the Yeast *Debaryomyces hansenii*." *J. gen. Microbiol. 89*, 256–264.

Kulaev, I. S. 1975. "Biochemistry of Inorganic Polyphosphates." *Rev. Physiol. Biochem. Pharmacol. 73*, 131–158.

Kwon-Chung, K. J. 1975. "A New Genus, *Filobasidiella*, the Perfect State of *Cryptococcus neoformans*." *Mycologia 67*, 1197–1200.

MacDonald, J. C. 1965. "Biosynthesis of Pulcherriminic Acid," *Biochem. J. 96*, 533–538.

Marchant, R., and D. G. Smith. 1968. "Bud Formation in *Saccharomyces cerevisiae* and a Comparison with the Mechanism of Cell Division in other Yeasts." *J. gen. Microbiol. 53*, 163–170.

McCully, E. K., and C. F. Robinow. 1971. "Mitosis in the Fission Yeast *Schizosaccharomyces pombe*: A Comparative Study of Light and Electron Microscopy." *J. Cell Sci. 9*, 475–507.

———. 1972. "Mitosis in Heterobasidiomycetous Yeasts." Parts I and II. *J. Cell Sci. 10*, 857–881; *11*, 1–31.

Mrak, E. M., and H. J. Phaff. 1948. "Yeasts," *Ann. Rev. Microbiol. 2*, 1–46.

Nakase, T. 1972. "Significance of DNA Base Composition in the Classification of Yeasts and Yeast-like Fungi." In *Fermentation Technology Today*. G. Terui, ed. 785–

791. Published by the Society of Fermentation Technology, Osaka, Japan.

Nečas, O. 1971. "Cell Wall Synthesis in Yeast Protoplasts." *Bacteriol. Rev. 35*, 149–170.

Phaff, H. J. 1977. "Enzymatic Yeast Cell Wall Degradation." In *Food Proteins*—Improvement through Chemical and Enzymatic Modification. R. E. Feeney and J. R. Whitaker, eds. *Advances in Chemistry Series 160*. American Chemical Society, Washington, D.C.

Phaff, H. J., and E. M. Mrak. 1948–1949. "Sporulation in Yeasts." Parts I and II. *Wallerstein Laboratory Communications 11*, 261–279, and *12*, 29–44.

Poon, N. H., and A. W. Day. 1975. "Fungal Fimbriae. I. Structure, Origin and Synthesis." *Can. J. Microbiol. 21*, 537–546.

Rattray, J. B. M., A. Schibeci, and D. K. Kidby. 1975. "Lipids of Yeasts." *Bacteriol. Rev. 39*, 197–231.

Rodrigues de Miranda, L. 1972. "*Filobasidium capsuligenum* nov. comb." *Antonie van Leeuwenhoek 38*, 91–99.

Spencer, J. F. T., and H. R. Sallans. 1956. "Production of Polyhydric Alcohols by Osmophilic Yeasts." *Can. J. Microbiol. 2*, 72–79.

Stewart, G. G. 1974. "Some Thoughts on the Microbiological Aspects of Brewing and Other Industries Utilizing Yeast." *Adv. Appl. Microbiol. 17*, 233–264.

Stodola, F. H., M. H. Deinema, and J. F. T. Spencer. 1967. "Extracellular Lipids of Yeasts." *Bacteriol. Rev. 31*, 194–213.

Tanaka, T., H. Kita and K. Narita. 1977. "Purification and Its Structure-Activity Relationship of Mating Factor of *Saccharomyces cerevisiae*." *Proc. Japan Acad. 53*, 67–70.

Tingle, M., A. J. Singh Klar, S. A. Henry, and H. O. Halvorson. 1973. "Ascospore Formation in Yeast." XXIII, *Symp. Soc. Gen. Microbiol.* Microbial Differentiation, 209–243.

Työrinoja, K., T. Nurminen, and H. Suomalainen. 1974. "The Cell Envelope Glycolipids of Baker's Yeast." *Biochem. J. 141,* 133–139.

von Arx, J. A. 1972. "On *Endomyces, Endomycopsis* and Related Yeast-like Fungi." *Antonie van Leeuwenhoek 38,* 289–309.

Webb, A. D. 1972. "Volatile Aroma Components of Wines and Other Fermented Beverages." *Adv. Appl. Microbiol. 15,* 75–146.

Webb, A. D., and J. L. Ingraham. 1963. "Fusel Oil," *Adv. Appl. Microbiol. 5,* 317–353.

Wickerham, L. J. 1951. "Taxonomy of Yeasts," *U. S. Dept. Agriculture, Techn. Bull. No.* 1029, 1–56.

Wickerham, L. J., and K. A. Burton, 1962. "Phylogeny and Biochemistry of the Genus *Hansenula,*" *Bacteriol. Rev., 26,* 382–397.

Wickner, R. B. 1976. "Killer of *Saccharomyces cerevisiae:* A Double-stranded Ribonucleic Acid Plasmid." *Bacteriol. Rev. 40,* 757–773.

Yamada, Y., and K. Kondo. 1972. "Taxonomic Significance of the Coenzyme Q System in Yeast and Yeast-like Fungi." In *Fermentation Technology Today.* G. Terui, ed. 781–784. Published by the Society of Fermentation Technology, Osaka, Japan.

Yamada, Y., M. Arimoto, and K. Kondo. 1977. "Coenzyme Q System in the Classification of Some Ascosporogenous Yeast Genera in the Families Saccharomycetaceae and Spermophthoraceae." *Antonie van Leeuwenhoek 43,* 65–71.

Index